Population Biology
Retrospect and Prospect

Population Biology

Retrospect and Prospect

Edited by
CHARLES E. KING
PETER S. DAWSON

COLUMBIA UNIVERSITY PRESS
New York 1983

Library of Congress Cataloging in Publication Data
Main entry under title:

Population biology.

Proceedings of the 41st Annual Biology Colloquim, held at Oregon State University, Apr. 25-26, 1980.
Bibliography: p.
Includes index.
1. Population biology—Congresses. I. King, Charles Everett, 1934- . II. Dawson, Peter S., 1939- . III. Biology Colloquium (41st: 1980: Oregon State University)
QH352.P622   1983   575.1'5   82-17762
ISBN 0-231-05252-9

Columbia University Press
New York   Guildford, Surrey

Copyright © 1983 Columbia University Press
All rights reserved
Printed in the United States of America

Clothbound editions of Columbia University Press books are Smyth-sewn and printed on permanent and durable acid-free paper.

Contents

	Introduction	vii
1.	The Environmental Component of Evolutionary Biology Douglas E. Gill, Keith A. Berven, and Beverly A. Mock	1
2.	Clonal Diversity in Cladoceran Populations Paul D. N. Hebert	37
3.	The Extent and Consequences of Heritable Variation for Fitness Characters Conrad A. Istock	61
4.	Measuring Genetic Variation in Natural Populations: Where Are We? Jeffrey R. Powell	97
5.	The Biochemical and Physiological Bases of Aminopeptidase-I (Lap) Polymorphism in *Mytilus edulis* Richard K. Koehn	117
6.	Genetic Variation, Environmental Heterogeneity, and Evolutionary Stability Peter S. Dawson and Russel A. Riddle	147
7.	Plant Parentage: An Alternate View of the Breeding Structure of Populations Donald A. Levin	171
8.	Achieving Synthesis in Population Biology Wyatt W. Anderson	189

Biology Colloquium: Roundtable Discussion   205
    Wyatt W. Anderson
    Peter S. Dawson
    Douglas E. Gill
    Paul D. N. Hebert
    Conrad A. Istock
    Charles E. King
    Richard K. Koehn
    Donald A. Levin
    Jeffrey R. Powell

Author Index   225

Subject Index   233

List of Contributors   235

Introduction: A Fable

Being a Historical Note on the Origin of Population Biology

Many years ago, when the world was still uncrowded and covered with trees and grass, there lived a little boy whose name was Scientist. Like many young children, Scientist spent most of his time poking about and asking questions. He wondered why thunder was so loud and why some rocks were pink and others grey, why lemons were bitter and why, after a rain, there were worms on the ground where before there had been none. Most children, when they grow up, know all the answers. The fact that they once wondered why bees visit flowers is treated as a distant embarrassment, sort of like not being able to tie one's own shoelaces. But Scientist was different; when he grew up he kept asking questions and trying to find answers. As the years passed, he had children and taught them his questioning ways. Four of these children were called Physicist, Geologist, Chemist, and Biologist. They also asked many questions, but they asked them about different things and each followed a different path.

All of Scientist's children had many children of their own, and their children's children had still more children. There were *lots* of children. A few years ago, about the time Evolutionist was born, some of Biologist's great-great-grandchildren decided to have a family reunion. But Embryologist was too busy cracking eggs and Taxonomist had just found a lot of animals and plants without Latin names, so neither of them could go. The same thing happened with the others—Physiologist, Ecologist, Geneticist, and Anatomist—all of them were much too busy to go. Besides, it didn't sound like much fun so they all made excuses and the reunion was called off.

At about the same time, tragedy struck: the world ran out of new one-word names for children. All the children who were born after that had to be given two- or three-word names like "Comparative Vertebrate Embryologist," "Seed Plant Taxonomist," "Population Geneticist," and "Population Ecologist." And another tragedy is that all these children wondered about different things. So after a while, the only time the members of a family ever talked to each other was when they sat down at the table to eat and someone couldn't reach the salt.

But fortunately the story does not end here. A truly wondrous event happened a few years ago while one of the Biologist's great-great-great-grandchildren was poking about in the winter woods. The handsome lad had been in these woods many times before and every question he could think of had also been asked many times before. Then he chanced upon a pretty little girl who was grinding up some minerals to see what they were made of. Now it was a cold, dismal day and the girl was also rather bored so the two children began to talk. After a while the girl, one of Chemist's great-great-great-grandchildren, asked the boy what kind of chemicals animals and plants were made of. The boy couldn't tell her. He'd never asked that question so after they talked about it for awhile they decided to find out. For the rest of that winter he asked her questions about chemistry and she asked him questions about biology. They had many long talks and found that they had really been interested in many of the same things all the time. As the seasons passed they made many new discoveries and had a child of their own. That child was named Biochemist. And, as you may know, Biochemist grew up to be honored and noble—prized by all.

Now, another wondrous thing happened just a few years ago (in 1978 at Syracuse University, New York) when Population

Historical Note

Geneticist and Population Ecologist chanced upon each other while trying to escape the boredom of their tired old questions.

The roster of speakers at the Syracuse Population Biology Symposium included many of the eminent population geneticists and ecologists of the day. Obviously, the intellectual climate was prepared for something important to happen. However, the formal sessions were marked more by ecologists talking Ecologese and geneticists speaking Genetician than by effective communication on shared problems. Perhaps the difficulty stemmed from the absence of a common language.

In contrast, the informal evening sessions, spontaneously held at the local pubs and bistros, provided a much different environment. Here the participants and discussions were far less inhibited and, perhaps because few were speaking clearly anyhow, the lack of a common lexicon was viewed as a minor inconvenience rather than as a major obstruction to shared thought.

A direct consequence of the Syracuse gathering was the rapid institution of courses in Population Biology at many of the leading colleges and universities in the United States. As these courses matured and the field developed, language barriers toppled; it became accepted that biologists who studied populations were to have familiarity with the major concepts of both population genetics and population ecology.

Our major purpose in organizing this colloquium was to evaluate the state of research in the areas of union between population genetics and ecology. The best way to judge this union, we concluded, was to gather a group of empiricists and ask them to present a synopsis of their current research. Although theory provides one type of window on the state of a discipline, there has never been any shortage of concepts in population biology and it is the empirical base that limits development in this area of scholarship. We make no attempt herein to draw evaluative conclusions for we feel the included papers are an eloquent statement of the current state of population biology.

A final hope we had in organizing this meeting was to take advantage of the gathered expertise to assess the directions that will, or at least should, be followed in the next decade. The vehicle we chose was to designate one of the speakers, Wyatt Anderson, to give a personal assessment of the field and then to lead a round-table discussion of all participants on future developments. Anderson's paper and the round-table discussion bring the present volume to a close. Only time will permit a satisfactory evaluation of the success or failure of this part of the program.

The Biology Colloquium has been held on the Oregon State University campus annually since 1939 with the exception of four years during World War II. The 1980 meeting was the forty-first Annual Biology Colloquium and it could not have taken place without the financial support and intellectual committment of various academic schools and departments of Oregon State University, notably the College of Science and Department of Zoology. Additional support was contributed by the U.S. Environmental Protection Agency, Phi Kappa Phi, and the Sigma Xi Scientific Research Society. Ms. Eva Millemann participated in all phases of the Colloquium organization from arranging travel and accommodations to proof reading the final manuscripts. We gratefully acknowledge her numerous and indispensable contributions. A note of gratitude is also due Mr. Leslie Bialler of Columbia University Press. Without his attentive eye and expeditious editing efforts, we suspect our final manuscript would still be on a clandestine shelf gathering Gotham City dust.

CHARLES E. KING
PETER S. DAWSON
Oregon State University
Corvallis, Oregon

Population Biology
Retrospect and Prospect

1 The Environmental Component of Evolutionary Biology

DOUGLAS E. GILL, KEITH A. BERVEN, AND BEVERLY A. MOCK

A major controversy during the last generation of population biologists has been the adaptive significance of life history traits and reproductive patterns in organisms (see Stearns 1976, 1977). Included in this discussion have been arguments about the advantages to individual fitness of semelparous and interoparous reproduction (Cole 1954; Bryant 1971; Gadgil and Bossert 1970; Charnov and Schaffer 1973), age and size of maturity (Lewontin 1965; Schaffer 1974; Wilbur, Tinkle, and Collins 1974; Fagen 1972), the conflict between r and K selection (Pianka 1970, 1972; Hairston, Tinkle, and Wilbur 1970; Pianka and Parker 1975; Gill 1978d), evolution of breeding systems (Brown 1964; Emlen and Oring 1977), and the like. To be sure, Lack (1954) initiated the debate in his celebrated hypothesis that clutch size in birds was carefully molded by natural selection to an optimal size set by the food-carrying constraints of the parents. The polemics have never been concerned with *whether* the traits are adaptive, but rather just *how* are they adaptive for various environments (Murphy 1968).

The very choice of words (namely, the evolutionary basis of life history traits) reveals that the polemicists assume that the traits under discussion are heritable and subject to the laws of natural selection. Of course, at some epistemological limit, every biological characteristic has a genetic origin and therefore is potentially modifiable by natural selection. Yet, any claim that the "special trait X has as adaptive significance" is shorthand for saying that trait X is inherited, has had genetic variability, has had a history of manipulation by natural selection, and is currently in a character state conferring optimum functionality in a relevant environment.

The question of concern here is whether the observed patterns in life history traits and reproductive patterns have a genetic basis and are the results of natural selection, or whether most of the observed patterns are phenotypic, either caused directly by the current environment or indirectly by environmental effects during development. Surprisingly few efforts have been made to separate genetic effects from local environmental effects in natural populations. The classic study of Clausen, Keck, and Hiesey (1940) on geographic variation in plant morphology and reproduction still stands as a lonely model. The heritability (the portion of observed variation in a trait attributable to genetic variation) of many life history traits in laboratory culture of *Drosophila* spp., livestock, agricultural crops, and some natural populations has been estimated and found to be uniformly low (Istock, article 3). Although such data indicate that most of the observed variation is due to direct environmental effects, they also support the premise that natural selection has weeded out most of the genetic variability associated with the trait, that optimal genotypes now dominate the population, and that there is a high degree of genetic determinism.

There is no question that the potential power of natural selection to mold any trait into an optimal state—whether biochemical, morphological, ecological, or behavioral—is great. Elementary models of selection on polymorphic loci in which the genotypes have unequal fitnesess show that equilibrium frequencies can be attained in a short time with surprisingly modest selection differentials. These models assume, of course, that the selection coefficients remain constant for many generations and by implication that the environment in which the population resides is unchanging (or at least is predictable).

To the degree that natural environmental conditions vary stochastically at the local level, the actual power of natural selection to establish optimum genotypes at fixed equilibria may be greatly diminished. Unpredictable fluctuations in weather, critical resources, and surrounding species composition do occur in the lives of most organisms. Such vagaries of the environment can have catastrophic effects on the distribution and abundance of populations in nature, even causing local extinctions (Andrewartha and Birch 1954; Birch 1957). In a chaotic environment, it may be very difficult to identify either relevant genetic equilibria or the fitness of character states which do become fixed. For example, if an environment fluctuates between extreme temperatures at a rapid rate, one can expect any of three outcomes: (1) net selective differentials between extreme genotypes of zero, producing effectively "neutral" alleles at loci coding for temperature sensitive enzymes; (2) fixation of one genotype due either to chance in a random walk or to intense selection during a prolonged spell at one extreme; (3) maintenance of alternative genotypes in heterotic polymorphism. Depending on the nature of the variability (stochastic, oscillatory, predictable, etc.) the actual power of selection may be either negligibly small or greatly amplified; at any rate it is probably very different in real life than its potential in a stationary environment (MacArthur 1976; Wiens 1977).

There have been efforts on the part of theoretical population biologists to model the process of natural selection and population density regulation in varying environments. In 1971 it was popular to develop models of selection within the context of density-dependent growth (King and Anderson 1971; Charlesworth 1971; Roughgarden 1971). Selection in populations distributed across patchy environments has also been considered theoretically (MacArthur and Pianka 1966; MacArthur and Wilson 1967; Maynard Smith 1970, Roughgarden 1974; Wiens 1976 for review). Increasingly sophisticated analyses of selection in different kinds of variable environments are under investigation (May 1973; see ch. 20 in Rougharden 1979).

The purpose of this article is to illustrate the kinds and magnitudes of environmental variabilities to which organisms are subject in natural settings. The vagaries of the environment may influence the breeding frequency of individuals, determine the reproductive success of populations in patchy breeding habitats, and directly affect such important life history characteristics as age and size of first reproduction. Three examples are drawn from long-term field studies done in our laboratory. In the first example, the influence of variable levels of parasitism on the breeding frequency of the red-spotted newt, *Notophthalmus viridescens* (Rafinesque), is suggested. In the second example, the patterns of reproductive success of a wide variety of organisms, including frogs, salamanders, and dragonflies, in a series of woodland ponds are presented and discussed. In the third example, the environmental and genetic components of clinal variation in life history characters in frogs are separated by elegant field and laboratory experiments. The take-home message is that many of the character states we observe in the field are the direct result of environmental circumstances during ontogeny and breeding; they may not be genetic adaptations in the situations we see them.

Population Dynamics and Breeding Frequency of the Red-Spotted Newt

In a series of previous publications, we reported the results of a three-year study on the population dynamics of a metapopulation of the red-spotted newt, *N. viridescens* (Gill 1978a, 1978b, 1978c, 1979). After six years of study the results have confirmed many of the earlier interpretations but have altered several others. Breeding population size in each pond continues to show constancy over time (figure 1.1). In fact, four of the five ponds appear to be equilibrating at a population size of 200 individuals. Whether this is a true saturation density for all ponds of this size remains to be seen.

Turnover rates of adults to new recruits continue at a high level: about 40 percent in males and about 50 percent in females.

Environmental Component

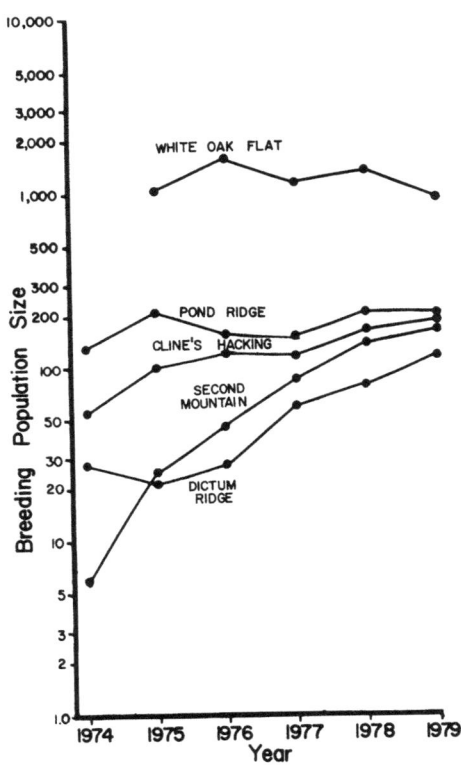

Figure 1.1. Sizes of five breeding populations of the red-spotted newt in the Shenandoah Mountains, Virginia.

These estimates are substantially lower than the previous estimates of 60 percent because of two patterns revealed by the long-term study. The first is that the drift fences surrounding each pond are imperfect and fail to intercept some individual newts as they arrive in the spring or depart in the fall. Whereas the majority of adult records show a consistent pattern of entrances and exits, the leakiness of the fence has created some inconsistencies or gaps in the records of some individuals (table 1.1). It is evident in the record of female P2, for example, that she entered Pond Ridge Pond in the spring of 1975 and the spring of 1977 and departed in

Table 1.1 Representative Recapture Records of Two Adult Female Red-Spotted Newts, *Notophthalmus viridescens* Rafinesque At a Pond, in the George Washington National Forest, Virginia.[a]

	Female P2		Female P5	
Original Capture	3 August 1974	OUT	3 August 1974	OUT
Recapture	15 August 1975	OUT(!)	no 1975 record	
Recapture	28 March 1976	IN	no 1976 record	
Recapture	16 September 1976	OUT		
Recapture	8 September 1977	OUT(!)	24 June 1977	IN
			1 September 1977	OUT
Recapture	6 April 1978	IN	30 March 1978	IN
Recapture	16 November 1978	OUT		
Recapture	2 April 1979	IN	no 1979 record	
Recapture	31 March 1980	IN(!)	no 1980 record	

[a] Both individuals have been followed for seven years and have appeared at only Pond Ridge Pond. Female P2 illustrates the pattern of most females, which consistently return to breed every year. Instead of a perfect recapture record (in which a Spring IN and an Autumn OUT would show every year) P2 trespassed (!) the drift fence three times because of fence leakiness. Female P5 illustrates one pattern of activity in which females skip a variable number of breeding seasons. P5 was not present at any other pond (possible breeding site) in 1975 and 1976; she was presumably terrestrial for two years. She has not been seen since 1978.

the fall of 1979 without being intercepted and recorded. The rate of trespassing due to fence leakiness is low, averaging about 5 percent per year, but is variable among ponds. Analysis shows that virtually all the gaps in the records of adult male newts can be explained by fence leakiness. A full discussion of how this error of technique alters interpretation of demographic and life history patterns is being prepared.

The second reason why the original estimates of population turnover rates were too high is the tendency of some females to skip breeding seasons. Males clearly breed every year, but the records of females show many gaps of long duration. Whereas the majority of females have records that resemble that of P2, many have records that are filled with gaps such as in P5 (table 1.1). These inconsistencies are more than can be accounted for by compounding the probabilities of seasonal trespass rates. Either females are more capable of trespassing fences than males, or there is a genuine tendency to skip breeding opportunities for variable lengths of time. Of known trespasses within seasons, there is no greater tendency for females to trespass than males. Thus, we interpret the large inconsistencies in the records of females as genuine skips. A full discussion of how the reappearance of females after periods of absences has altered the interpretation of demographic variables, especially age-specific survivorship, is being prepared.

Of interest here is whether the variable frequency of breeding in female red-spotted newts has a genetic basis and is an "adaptive strategy" dealing with some feature of their environment, or whether it is directly caused by environmental circumstances befalling some unfortunate few. We discount the first possibility because the pattern of skipping does not relate to the boom-or-bust reproductive patterns in the ponds (see below). Instead we have data suggesting the second possibility. A likely candidate for the unfortunate circumstances is endoparasitism by trypanosomiasis.

For the past three years we have studied the pattern of infection of trypanosomiasis in these marked populations of the red-spotted newt. We have taken advantage of the unusual opportunity afforded by the established, marked host populations of newts to study the interactions of host and parasite population dynamics. The interactions appear to be mutually regulatory. In brief, the infecting parasite, *Trypanosoma diemyctyli* (Tobey) is a flagellated protozoan transmitted to adult newts by the amphibian leech *Batrachobdella picta* Verrill. In the case of the newt trypanosomiasis, we have established the pattern of age-specific incidence, prevalence, and intensity. We shall report elsewhere the details of these results and our conclusions about the several mechanisms of suprapopulation density regulation in *T. diemyctyli* (Mock and Gill, in prep.; Mock, in prep.).

Here we draw attention to the data that relate breeding frequency of adult newts with their intensity of their infection by trypanosomiasis. By August, the majority of adult newts were infected, but the intensity of infection in older newts was significantly less than the intensity in young adults (figure 1.2). Of startling interest was the additional significant variation in levels of infection among old adult newts of varying breeding histories. Among the 3-year-old adults, some individuals were known to

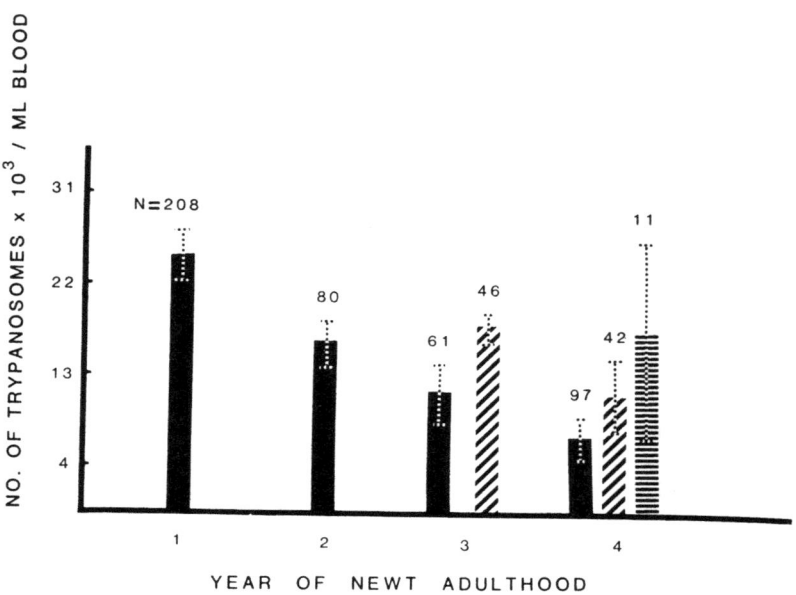

Figure 1.2. Intensity of infection of trypanosomiasis in adult red-spotted newts according to adult age.

Note: Intensity of infection is measured as the number of trypanosomes counted in a known volume of blood take from the caudal artery. Samples were taken in August and September, 1978. *Solid bars:* individuals that bred in ponds every year of their adulthood. *Diagonal striped bars:* individuals that skipped one year of breeding. *Horizontal striped bars:* individuals that skipped two years of breeding. Numbers above bars are sample sizes. Error bars are two standard deviations of the mean.

have been present in breeding ponds all three years and others that skipped one (the middle) year. Those that skipped one year showed significantly higher levels of infection than those that returned every year. Similarly, among the 4-year-old adults, some individuals were present all four years, others missed one year (skipped either the 2d or 3d year), still others missed two years (skipped both the 2d and 3d year). Again, those that were present all four years had the lowest density of infection. Level of infection increased with the number of years skipped; therefore, individuals that skipped two years had significantly higher levels of infection than those that returned every year.

At the very least, these data suggest that those individuals with

Environmental Component

gaps in their records are biologically distinguishable from those that show consistent records of breeding every year. They provide the strongest evidence that large gaps in the records of females are genuine skips in breeding activity rather than trespasses due to a leaky fence.

We require an explanation for the correlation of levels of infection and tendency to skip breeding opportunities. We hypothesize that the relationship may be casual. We have already noted that young adults, those breeding for the first time, acquired the heaviest infections by the end of their first summer. Yet there was variation in levels of infection among individuals within this even-aged cohort; most individuals were comparatively lightly infected while a small minority carried very heavy infections (Mock and Gill, in prep; Mock, in prep.). We suggest the females that acquire exceptionally heavy levels of infection in their first adult year are debilitated in a way that prevents them from breeding in subsequent years. The heavier the initial infection, the more years are missed until the next time of healthy sexuality. The majority of females with modest infections are capable of breeding in every subsequent year.

The hypothesis supposes that the debilitating effects of trypanosomiasis are greater on the breeding condition of females than they are on males because of the greater nutrient and energy demands of egg production than spermatophore formation. Of course, we have no direct evidence that bears on the energetic differentials of female and male breeding. Nevertheless, it is noteworthy that the trypanosomiasis did not influence physiological or ecological traits that would be associated with annual survivorship of adult newts. Not only was there no significant difference between male and female survivorship, but also the rate of mortality was constant across adult ages (Type II) for both sexes (figure 1.3). This constant rate of mortality irrespective of adult age is not consistent with the significant decline in levels of infection with adult age. Thus, trypanosomiasis in the red-spotted newt does not affect probability of adult survival but may influence frequency of breeding in adult female newts.

We are specifically testing the hypothesis that trypanosomiasis

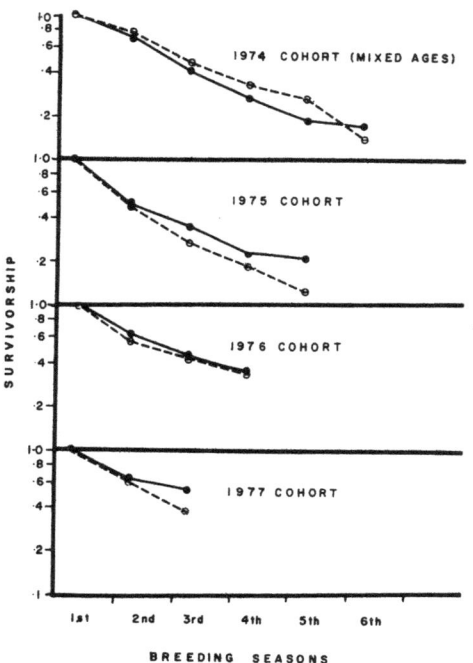

Figure 1.3. Age-specific survivorship of adult red-spotted newts in the Shenandoah Mountains.

Note: Data are pooled from four ponds (PR, CH, DR, and SW) which have been studied since 1974. Each point is the proportion of the original cohort known to be alive in the *i*th breeding season. *Solid lines and dots:* males. *Broken lines and open circles:* females.

causes irregularities in frequency of breeding in female newts by predicting the pattern of future breeding in the 1980 cohort of new adult newts. By screening this cohort of adults for trypanosomiasis in the spring of 1980, we predict the individuals that acquire the heaviest levels of infection by the end of the summer of 1980 will be those that will miss subsequent breeding opportunities.

Why is this interaction between trypanosomiasis and frequency of breeding in female newts relevant to the main point? Obviously, parasitism is a trait acquired during the life of the newt. The level of infection acquired is largely determined by stochastic events that befall individuals during their breeding activities. Individual newts have little control over the number of leeches attacking them or the number of trypanosomes injected into their bloodstreams. We have evidence that levels of infection are related to attack rates (Mock and Gill, in prep.). If we are correct in hypothesizing that levels of parasitic infections influence breeding frequency in newts, and that levels of infection are acquired characteristics, then it would be a mistake to speculate on the "adaptive significance" of observed fecundity schedules of the red-spotted newt.

Levels of infection are also influenced by the immunological responses of hosts to infecting parasites. Efficiency of immunological response or immunological competence can have a genetic basis. Thus, individuals with exceptionally low levels of infection may be genetically predisposed to immunity. Hence, some traits controlling the impact of parasitism can evolve through natural selection. This observation, discussed elsewhere (Mock, in preparation; Mock and Gill, in preparation), does not detract from the significance of our findings that acquired characteristics directly influence life history traits.

Reproductive Failure in the Red-Spotted Newt

The astonishingly erratic reproductive performance of red-spotted newts in these ponds has continued for six years (figure 1.4). Emergence of young efts in White Oak Flat Pond (WOF) peaked explosively in 1976 and has declined steadily ever since. Pond Ridge Pond (PR) produced nearly nothing for four years in a row and then suddenly became bountiful in 1978 and continued to be productive in 1979. Similarly, Cline's Hacking Pond (CH) was a dismal producer for five years before bringing forth over a thousand juveniles in 1979. Dictum Ridge Pond (DR) had only one impressive peak (1978) after four years of nothing. Altogether, there have been five spectacular booms in reproduction in the 29 pond-years of study, and four of the five ponds have had their red letter days.

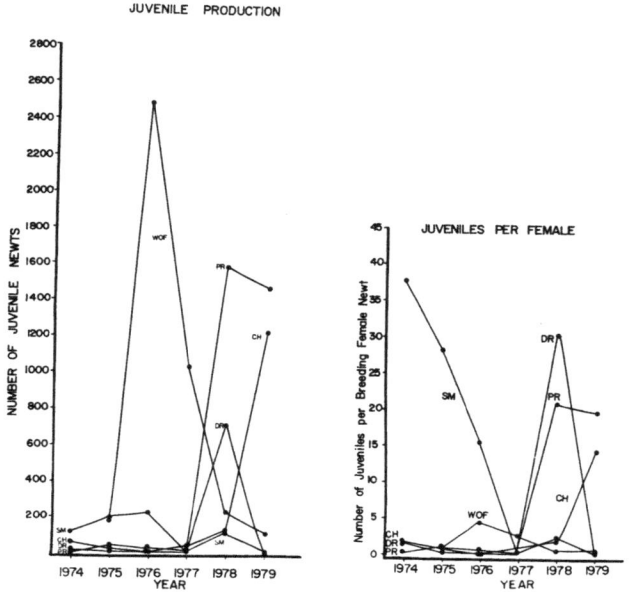

Figure 1.4. Production of post-metamorphic newts, called juvenile efts, from five ponds over six years.

Left: Actual number of efts captured in pitfall traps adjacent to a complete drift fence positioned 1–2 meters from the pond edge. *Right:* the average number of juveniles per female produced in the six years. The total number of juvenile efts was divided by the number of breeding female newts.

When converted to production per adult female, the pattern remains as erratic, but seven high points are evident. Because of the small size of the breeding population in Second Mountain Pond (SM) in 1974, 1975, and 1976, the per capita reproduction in that pond was extraordinarily high compared to the other ponds. The magnificent peak in WOF in 1976 shrank to a small bump when the number of emerging juveniles was divided by the very large number of breeding females. The raw production peaks of PR, DR, and CH were preserved after being transformed to the per capita rate. These per capita figures are the direct estimate of the advantage to the average individual which happened to be in each pond.

The erratic nature of the juvenile production contrasts sharply with the dynamic constancy of the breeding populations in the five ponds. The stationary pattern of the population size, age structure, and sex ratio of adult newts at each pond does not relate to its pattern of wild fluctuations in juvenile production. As far as we can determine, all adults in each pond were healthy, sexually active, and fecund. Even though there was a high turnover of adults in the breeding stocks between successive years, there was also a 50 percent continuity by virtue of the return of previously breeding adults. Thus, the abrupt changes in juvenile production from bust to bloom (e.g., PR in 1977 and 1978) or vice versa cannot be ascribed to a significant alteration of genotypic frequencies in the breeding population. This would be true no matter how large the fitness differentials between genotypes, even if the fitnesses were inherited and invariable.

The best explanation of the two patterns is that adult newts are relatively resilient to variations in the pond environment but that larval newts are extremely sensitive to local environmental conditions in both time and space. Both abiotic and biotic factors are implicated. We are confident that the droughts of 1976 and 1977, were calamitous to amphibian survival in all but the deepest ponds (e.g., WOF). In 1977 PR, CH, DR, and SM all shrank to a soupy puddle by mid summer and larval deaths were everywhere evident.

In addition, we have circumstantial evidence that it was the biotic portion of the pond environment that determined whether a deme of newts successfully produced efts or not. The amphibian leeches appeared to play particularly important roles. We had been impressed by the relatively high reproductive output from the newest and youngest pond (SM), especially in terms of the number of juveniles per female, during the first three years of the study (1974–1976). At that time SM was the only pond in which no leeches had been observed. In 1977, not only the year of the drought but also the first year in which leeches were observed in SM, newt reproduction collapsed to zero. Leeches have been observed in that pond every year since 1977, and newt reproduction has continued to be low despite near normal patterns of rainfall. Thus qualitatively, good newt reproduction was associated with the absence of leeches and poor reproductive success was associated with their presence.

We had also noticed that the leeches were conspicuously less abundant in successful pond-years than they were in failing pond-years. Moreover, it was exciting to note that there had been an absence, or at least sharp reduction, in the number of breeding adult wood frogs, *Rana sylvatica*, in the pond-year preceding the explosive production of juvenile efts (e.g., in 1975 no wood frogs appeared at WOF in the spring). Together with knowledge of the breeding biology of the leech, *B. picta*, these two observations lead to an hypothesis relating the presence of wood frogs in one year to the failure of newt reproduction in the next year (figure 1.5).

As these ponds thaw in February, the leech populations are composed of large, gravid adults which readily attack any amphibian that swims nearby. Typically the first amphibians encountered are breeding wood frogs. The first blood meal taken in the year is

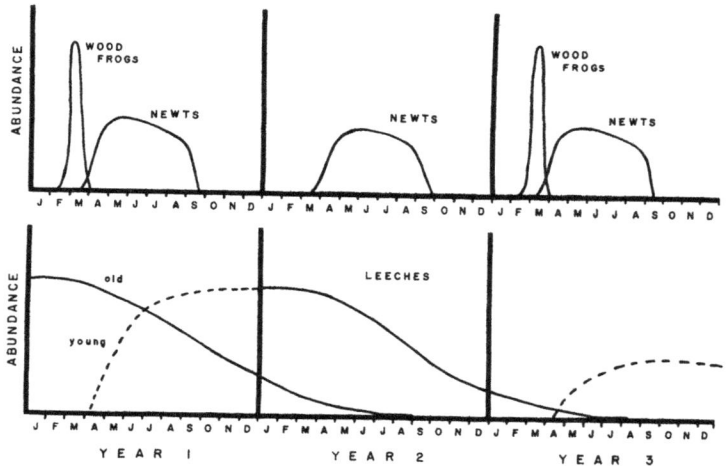

Figure 1.5. Schematic representation of the hypothetical role played by the breeding population of wood frogs on the eventual reproductive success of the red-spotted newt.

Note: In years that wood frogs pour into a pond for breeding in February, gravid leeches obtain a critical blood meal to complete egg and larval development. High population densities of leeches will sustain into the succeeding year. In years that wood frogs fail to appear, reproduction in the leeches is destroyed. A crash in the populatoin density of leeches ensues the following year. This relative absence of leeches frees the population of larval newts from a major predator. Large numbers of larvae can then successfully complete metamorphosis and emerge as juvenile efts.

critical to the mature leeches for the final maturation of the eggs *in utero* and for the sustenance of the adult leech as it broods its clutch during the month of March. By the time the leech eggs hatch, the wood frog adults have departed and the only abundant adult amphibians in the pond are red-spotted newts. During the months of May and June, both adult leeches and hundreds of young leeches attack and feed voraciously on adult newts. Although we have no direct observations to support the assertion, we presume that larval newts are also attacked at this time, because larval frogs (tadpoles) are also attacked by scores of leeches.

It appears that reproductive success of leeches in these ponds depended upon that first wood frog blood meal. We hypothesize that in the absence of wood frogs, the critical energy supply for egg development both before laying and during brooding is missed. Failure to obtain that blood meal could generate catastrophic losses of eggs and larvae, and precipitate a collapse in the leech population density in the following year. During the year of a leech population nadir, the rate of attack on newts would also be at a low point. The survival rate of larval newts would be correspondingly high, as would the emergence of metamorphosing efts. Thus, the presence of wood frog breeding choruses one year might ultimately control the reproductive success of a deme of red-spotted newts in the next year through their influence on leech reproduction.

This hypothesis is based on circumstantial evidence rather than hard data for two reasons. First, until recently we had no way of determining leech population sizes accurately. All we have are our subjective field observations that leech population densities vary greatly among ponds and among years. (We have now developed a standardized technique of gathering comparative data by "baiting" leeches with tethered frogs for set lengths of time.) Secondly, we have only 5 "booms" of newt reproduction out of 29 pond-years. As a consequence, our sample size of population-level data points is too small to perform meaningful correlation analysis of the potentially important variables. Nevertheless we are sufficiently attracted to these circumstantial pieces of evidence that we are currently testing the hypothesis by direct experimental manipulation of breeding wood frog populations. In March 1980 control ponds were allowed their normal populations of breeding wood frogs while at an experimental pond adult wood frogs were forcibly prevented from entering and contributing blood to the resident (currently large) population of amphibian leeches. A large population of juvenile efts emerged from the experimental pond in 1981 as predicted.

Reproductive Failure in Other Amphibians and in Dragonflies

The wild, unpredictable fluctuations in juvenile production, so vividly observed in the demes of red-spotted newts, have also been a conspicuous feature of reproduction in the other amphibian species and some insects that breed in these ponds. In fact, all species for which we have quantitative data showed extremely erratic patterns of reproductive success; no species has produced a constant or predictable number of young in every pond in every year. In figure 1.6, the pattern for the two species of *Ambystoma* salamanders, the Jefferson's salamander, *A. jeffersonianum*, and the spotted salamander., *A. maculatum*, are presented. The lack of concordance among ponds and years, and the drought-produced disaster of 1977 are obvious. Neither the proximal nor ultimate causes of these patterns in the *Ambystoma* salamanders are known to us yet.

The patterns of juvenile production in two frog species, the wood frog *Rana sylvatica* and *R. clamitans*, are presented in table 1.2. We have only qualitative observations for the years 1974 and 1975; ever since 1976 we have instituted a program of precise enumeration of emerging froglets. In both these species a major cause of mortality in adult, juvenile, and larval frogs was a sudden epidemic of a bacterial disease. The disease totally destroyed entire pond populations of anurans but showed no spatial or temporal pattern of appearance at specific ponds. The importance of these catastrophes on the dynamics of the frog populations will be

Environmental Component

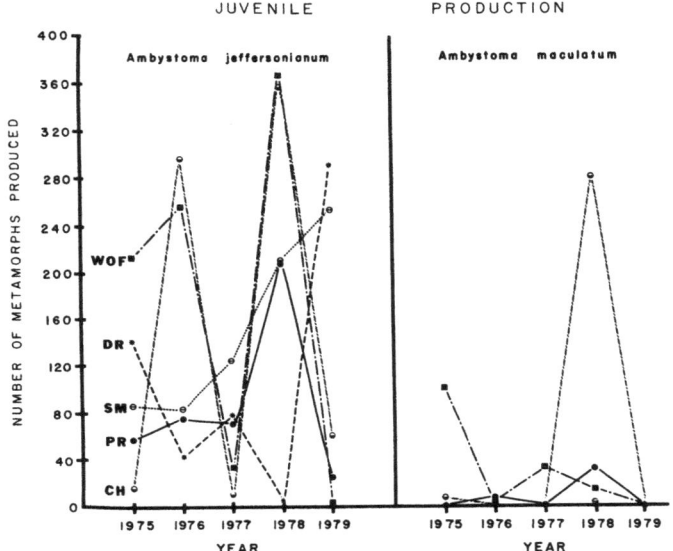

Figure 1.6. Juvenile production of juvenile *Ambystoma jeffersonianum* and *A. maculatum* from five mountain ponds.

Table 1.2. Juvenile Production of Frogs from Mountain Ponds.

	1974	1975	1976	1977	1978	1979
Wood Frogs (*Rana sylvatica*)						
Pond Ridge	1000's	100's	1593	682	314	192
Cline's Hacking	1000's	100's	6868	0	689	1656
Dictum Ridge	1000's	1000's	0	0	0	1823
Second Mountain	100's	100's	0	34	339	1963
White Oak Flat	100's	0	0	0	994	0
Green Frogs (*Rana clamitans*)						
Pond Ridge	0	0	0	0	0	0
Cline's Hacking	0	0	0	0	0	0
Dictum Ridge	0	0	0	0	0	0
Second Mountain	—	100's	122	0	0	0
White Oak Flat	—	100's	0	0	0	0

discussed in detail by K. Berven (in preparation). The point that we wish to stress here is that inconsistent patterns of juvenile production have been observed in anuran as well as urodele demes.

We have observed similar patterns of explosive juvenile production at particular ponds and/or in particular years in several insect populations. In 1976 and 1977 in pond PR only, there was an extraordinary production of the last instar of the predaceous water beetle, *Dytiscus verticalis*. Our best quantitative data have been collected by T. Halverson (unpublished) on the number of exuviae of six dragonfly species found on vegetation (and our drift fences!) surrounding each pond (figure 1.7). Numerous adults of each species have been seen at each pond. Yet, naiads of mostly one species emerged from each pond and the abundance of exuviae varied significantly between years. The causes of these pond-specific and year-specific patterns are currently under study.

Our data show that the pattern of juvenile production of species living in pond habitats is typically erratic and the variation in numbers spans several orders of magnitude. The erratic pattern is true of many species of organisms in as widely divergent taxa as amphibians and insects. These results suggest that widespread reproductive failure at the population level is a general phenomenon of species living in patch habitats. In our pond system at least, there is no way we as researchers can predict which pond in which year is going to have bountiful production. It is evident that the organisms themselves cannot either. The data available indicate that the causes of the variation in reproductive success are characteristics of the pond environment and are not in anyway attributable to the competence of the individual organisms.

Measurements of actual reproductive performance in the field do not provide the basis for interpreting the adaptive significance of the observed character states. It is necessary to show that the variation of a trait is inherited in order to justify speculation about its evolutionary importance. Our field studies have convinced us that the life histories ecologists commonly observe in nature are more consequences of local environmental conditions than they are expressions of carefully selected genotypes. It is an absolute requirement for future research in the area of the evolution of life history traits to perform critical experiments that separate the en-

Environmental Component

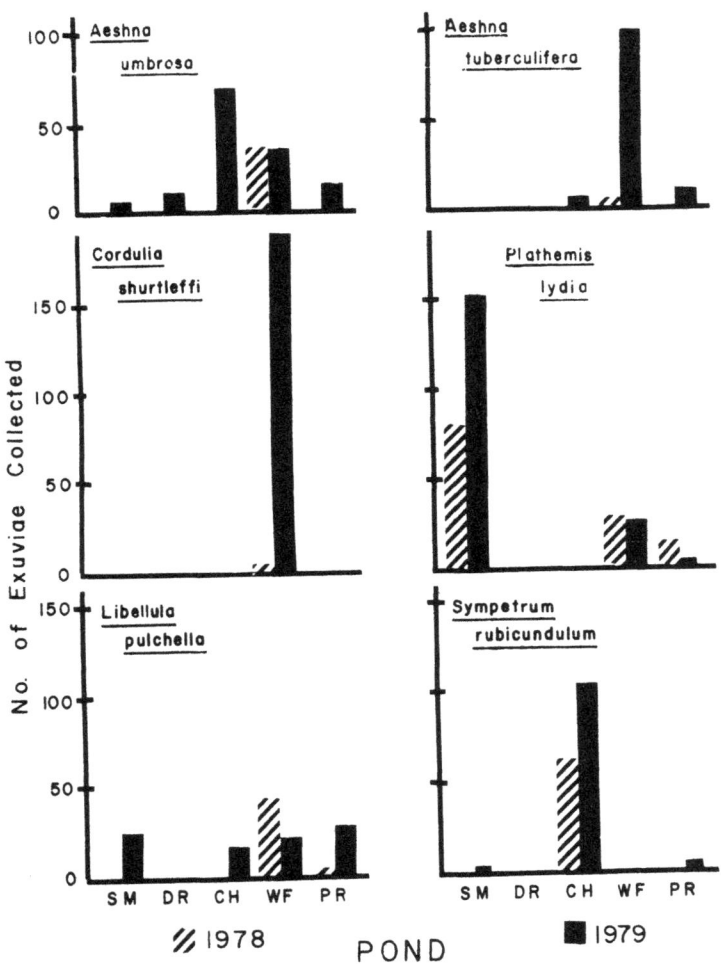

Figure 1.7. Collections of dragonfly exuviae at five mountain ponds in 1978 and 1979.

Note: Adults of six species have been observed at the ponds, but many species have failed to emerge or successfully metamorphose. Data from Halverson (in preparation).

vironmental from the genetic components of the observed traits. A third research project involving the experimental analysis of clinal variation in natural populations of frogs is directed toward that goal.

Genetic and Environmental Components of Life History Traits in Frogs

The Model

To illustrate the kind of experiments necessary to determine the adaptive significance of natural variation in life history traits, we refer to the model of clinal variation presented by Berven, Gill, and Smith-Gill (1979) (figure 1.8). Suppose a phenotypic trait (ϕ) varies systematically in a population across an environmental gradient which has convenient sampling positions 1–5. The smaller phenotype, observed at site 1, is referred to as $\phi 1$ and larger phenotype, observed at the other end of the gradient, is called $\phi 5$. It is desirable to perform an experiment in which developing individuals (g1) from site 1 are transplanted to the other sites and allowed to grow alongside resident individuals and to transplant individuals (g5) from site 5 to the other sites. Such a reciprocal transplantation experiment will show how much of the observed cline is genetic and how much is experimentally induced.

Five outcomes of these reciprocal transplantation experiments are possible: (1) If the phenotypes that emerge from the transplanted individuals coincide precisely with those of the resident individuals at each site, no genetic variation would be evident, the traits would show zero heritability, and the original cline would be interpreted as completely environmental. (2) If the transplanted individuals from site 1 retained their smaller phenotypes relative to resident individuals at all sites, and the g5 individuals retained their larger phenotypes at all other sites, then some portion of the original cline must have a genetic basis. The direction of the genetic differences would be in the same as the observed cline, so that the selection behind the genetic variation must have been co-gradient (figure 1.8B). If the slopes of the transplantation lines are zero, then the original cline was 100 percent genetic, and the phenotype traits would show 100 percent heritability. (3) If the g1 individuals became larger than any resident individuals and the g5 individuals developed smaller character states than any resident individuals at the experimental sites, (figure 1.8C), one

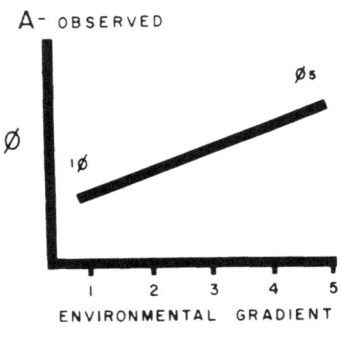
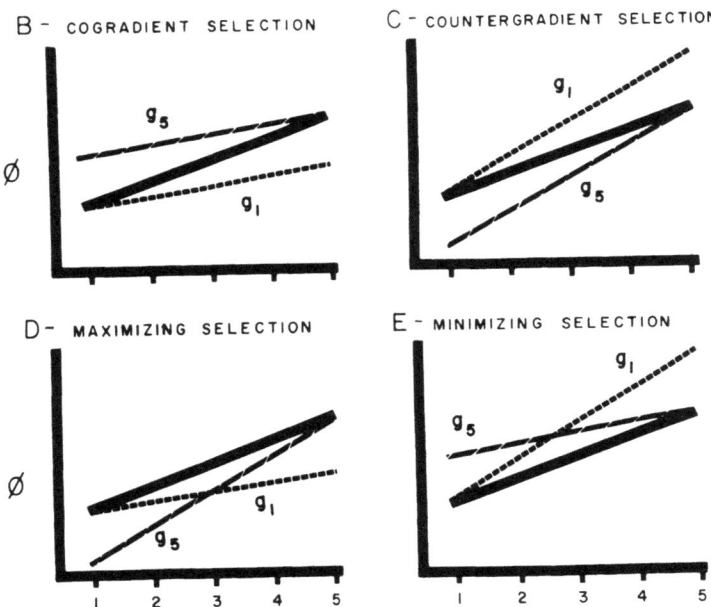

Figure 1.8. Model of clinal variation in an hypothetical species over an environmental gradient.

Note: $\phi 1$ refers to the phenotype observed at site 1 and $\phi 5$ refers to the phenotype observed at site 5. Diagonally striped lines indicate the possible phenotypes that may be expressed in individuals (g5) drawn from the genetic population at site 5 after they have been transplanted and cultured at the other sites on the gradient. The horizontally striped lines represent the phenotypes produced in individuals (gl) from population 1 following transplantation to the other sites. A: the natural cline. B: model of co-gradient selection, whereby experimental results show that the transplanted genotypes retain some of their original phenotypes. C: model of counter-gradient selection, whereby transplanted individuals show genetic tendencies antagonistic to the observed phenotypes at the original sites. D: model of maximizing selection, whereby all genotypes are induced by the environment to express phenotypes less than natural. E: model of minimizing selection, whereby all genotypes are induced by the environment to express phenotypes greater than natural. See text for detailed explanation. Model after that of Berven, Gill, and Smith-Gill (1979).

must conclude that the original cline had both an environmental and a genetic component. But the genetic variation is contrary to expectation: the innate, genetic tendencies are in the opposite direction of the observed character states. In effect, the genetic potentialities underlying the trait would allow a greater range of phenotypes than ever observed. Natural selection must have acted to minimize the magnitude of the inducible variation, and has acted counter to the observed cline.

Two other possible outcomes of the hypothetical transplantation experiment remain. (4) If individuals gl become smaller than any resident individuals, but g5 individuals retain their observed small character states when placed anywhere along the gradient (figure 1.8D), the observed cline would represent the maximum values of all the possible genetic tendencies. Environmental induction is evident because all individuals become large when placed at the high end of the gradient. Genetic variation is also evident but is both concordant (gl) and antagonistic (g5) to the observed cline. Natural selection has consistently selected the individuals with the maximum character states of the genetic potentialities. Finally, (5) if the gl individuals become larger than any resident individual, but g5 individuals retain their larger character states at all experimental sites (figure 1.8E), the observed cline would represent the minimum values of all possible genetic potentialities. As in the previous outcome both environmental and genetic components are evident in the natural cline, but natural selection has favored the minimum values of all innate tendencies.

The Experiment

Conspicuous clines in morphological, developmental, and life history traits were evident in both the green frog, *Rana clamitans*, and the wood frog, *R. sylvatica*, along an altitudinal transect from the Virginia coastal plain to the tops of the Shenandoah Mountains, 1000 m above sea level. Initial field studies on the green frog revealed that tadpoles in the lowlands were small in size and developed very rapidly, often completing metamorphosis within one season. In contrast, tadpoles from mountain ponds were large and developed very slowly, overwintering at least once and completing metamorphosis one or two years after oviposition (Berven et al. 1979). By culturing samples of eggs drawn from the extreme ends of the cline under identical conditions in the laboratory, we were able to dissect the portion of the observed variation in developmental and growth rates that was inducible by ambient environmental conditions and the portion that was innately associated with the population of origin.

The results of those experiments, reported in an earlier paper (Berven et al. 1979), were startling. Through manipulation of only temperature in the laboratory, we were able to reproduce the character states of the natural cline in the laboratory in both samples. However, there were significant differences between the two samples that could be attributed only to genetic differences between the two original ends of the natural cline. These genetic differences were not in the expected direction: when grown under cold temperatures that resembled the natural pond temperatures in the mountains, lowland tadpoles took longer to develop and reached larger sizes at metamorphic climax than the native mountain tadpoles. In the warm-temperature treatments, which were designed to resemble the natural temperatures of ponds at low elevations, the lowland tadpoles did develop as expected: they reached metamorphic climax earlier and at a smaller size than the mountain strains.

The interpretation of the results was inescapable. The observed developmental differences between highland and lowland populations were due mostly to environmental conditions (most probably water temperatures) in which the growing tadpoles found themselves. The large size and the slow development of green frog tadpoles in the mountains were not genetic traits *per se;* they were the results of ambient conditions that occurred during development. Where genetic tendencies were evident, natural selection had favored the fastest possible developmental rates (an example of maximizing selection of figure 1.8D) resulting in the smallest possible sizes at metamorphic climax (an example of minimizing selection of figure 1.8E), *given the ambient conditions*. Thus large sizes and slow developmental rates in the mountain green frogs do not have an adaptive significance in themselves.

It was possible to perform only laboratory experiments with larval green frogs; it was not possible to perform a genuine set of reciprocal transplantation experiments with the natural populations in the natural habitats. Such experiments were possible with another species, the wood frog, *R. sylvatica*. Clinal variation in the wood frog was similar to that observed in the green frog (table 1.3). Wood frog adults are very large in the mountains and small in the lowlands. Wood frogs reach maturity at a later chronological age in the mountains than in the lowlands. Both egg size and clutch size are larger in the mountains, and larval development takes longer there.

The experimental methods followed the desired design. First, lowland egg masses were transplanted to the mountains, and mountain egg masses were reciprocally transplanted to the low-

Table 1.3. Comparison of Morphological and Life History Traits of Mountain and Lowland Wood Frogs (1977-1978).[a]

	Lowland[b]	Mountain[c]
Adult Body Size (snout-ischium in mm)		
Males	41.7 ± 2.5	55.3 ± 3.1
Females	47.8 ± 3.7	64.4 ± 4.3
Age at First Reproduction (years)		
Males	1	3
Females	2	4
Mean Egg Number	642 ± 199	840 ± 215
Mean Egg Size (mm)	1.83 ± 0.28	2.28 ± 0.14
Larval Development		
Duration (days)	85	110
Size at Metamorphosis (snout-ischium in mm)	15.8 ± 1.6	18.7 ± 2.2

Source: Berven 1982 a,b.
[a]Mean ± 1 standard deviation.
[b]Coastal plain of Maryland (altitude: 43m).
[c]Top of Shenandoah Mountains (altitude: 1000m).

lands. Second, metamorphic juveniles from mountain populations were transplanted to lowlands, and reciprocally, metamorphic juveniles from the lowlands were transplanted to the mountains. And third, samples of eggs from both mountain and lowland sites were brought into the laboratory and cultured side-by-side under several temperature conditions. The latter experimental procedure was designed to provide direct comparison with the previous experiments with green frog tadpoles.

Our expectations that the features of environmental induction and genetic basis, especially the counter-gradient selection, of size at metamorphosis and developmental rates in wood frogs, proved true and false. In the first experiment, in which eggs were reciprocally transplanted between the lowlands and highlands and larval development was monitored, it was true that all tadpoles placed in the mountain environment took longer to complete metamorphosis, grew more per developmental stage, and reached a much larger size than those eggs placed in the lowland pond (figure 1.9). Indeed, the length of the larval period conformed strictly to ambient conditions and not to the population of origin. Thus a big, direct environmental influence on all traits under study was demonstrable.

However, eggs of mountain origin consistently had higher growth rates per developmental stage and climaxed at a significantly larger size in both environments than did individuals of lowland origin. Mountain tadpoles retained the larger sizes and faster, stage-specific growth rates characteristic of their natural populations, so where a genetic basis of larval developmental characters could be demonstrated, the direction was the same as that observed naturally; natural selection had acted co-gradiently, the opposite of its action in the green frog populations living across the same environmental gradient.

The second experiment, in which metamorphic juveniles were reciprocally transplanted between highland and lowland sites, separated the components of the size at and chronological age of sexual maturity in both males and females. The experiments permitted frogs of common parentage to spend their juvenile periods

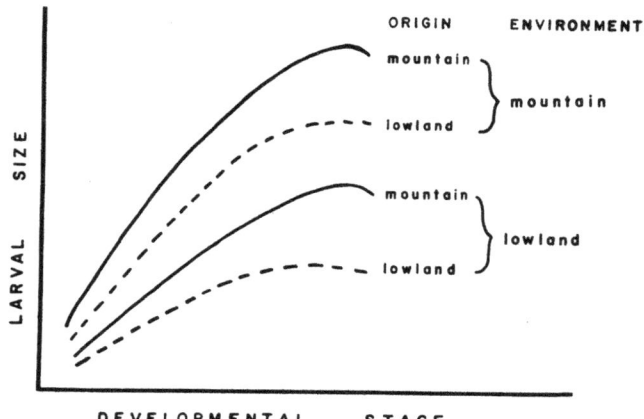

Figure 1.9. Larval development of wood frogs in experimental cages in natural ponds.

Note: Eggs were reciprocally transplanted between the ends of a cline over an altitudinal gradient. Growth per developmental stage is illustrated, whereby the sizes at metamorphic climax are illustrated for the four groups. Mountain tadpoles in mountain ponds and lowland tadpoles in lowland ponds are the controls. Tadpoles of mountain origin grew faster per developmental stage and attained a larger size at metamorphic climax. Not illustrated in this figure is the fact that all tadpoles in the mountain environment took a month longer to complete metamorphosis than did all tadpoles in the lowland environment. Figure is a facsimile of data from Berven 1982b.

in contrasting environments. Again, both environmental effects and genetic effects were significant. Whereas most males were 57mm in length at maturity in the mountains, they reached maturity at 52mm when transplanted to the lowlands (table 1.3). But this "transplanted" size was still 10mm larger than resident lowland males (42mm). Likewise, lowland males reached maturity at a larger size when transplanted to the mountains than did their siblings back in the lowlands, but their size was not as large as mountain residents. Exactly the same pattern of size at maturity was evident in females. Thus, there is a strong environmental effect during development on size at maturity. The genetic component was in the same direction as the natural cline; therefore selection had been co-gradient on size at maturity.

Age at maturity in males was the same story. Whereas most mountain males reached sexual maturity at three years of age (a

Environmental Component

few can do it in two years, some take four), the large majority of juvenile males transplanted to the lowlands took two years (a few made it in one year). In contrast resident lowland males typically matured in one year with a minority maturing in two years. Lowland juvenile males transplanted to the mountains took only two or three years to reach maturity. Thus there was a large environmental effect during development, such that all males took longer in the mountains than they did in the lowlands. But the genetic effect was co-gradient; wood frogs of lowland origin always reached maturity earlier than mountain wood frogs.

The determinants of the age of first reproduction in female wood frogs were also environmental and genetic. When lowland female juveniles were transplanted to the mountains, they took three years to reach maturity in contrast to the resident time of four years. In like manner, when mountain juvenile females were transplanted to the lowlands, they reached maturity at age two which was not significantly longer than the maturity time of resident lowland females. There were the size differences at noted above. Thus, the genetic component to age of first reproduction in female wood frogs was demonstrably co-gradient.

The juvenile transplantation experiment also illustrated how erroneous some interpretations of the evolutionary significance of clutch size and egg size variation can be when they are made from empirical evidence alone. In nature we observed that the large upland females produced significantly larger and 200 more eggs than lowland females (table 1.3). The unwary evolutionary biologist might speculate immediately on the adaptive significance of many, large eggs at high elevation. However, when the transplanted juveniles returned as breeding adults, we found that the lowland-transplants produced the same number of eggs as the mountain residents, and the clutch size of the mountain transplants nearly matched the clutch size of the average lowland resident (figure 1.10, top). According to our model (figure 1.8a), these results point to an alternative conclusion, namely that the observed phenotypic cline in egg number was actually environmental.

But these two conclusions fail to take into account a more fundamental aspect of the determinants of clutch sizes. The number of eggs produced by a female frog is functionally dependent upon her body size; larger females produce more eggs. The rates of change of egg number with female body size were the same in both the lowland and mountain populations, but the elevations (adjusted means) of the two regressions were significantly different (figure 1.10, top; Berven [1982a]. Thus, for a given body size, mountain females actually produced fewer eggs than the average lowland female. In fact, the transplantation experiment revealed that the clutch-size–female-body-size function of each population was an inherent, genetic relationship. Although the lowland transplants were environmentally induced to larger body sizes, their clutch sizes were simple extensions of the lowland function; they did not transfer to the mountain function. Similarly, the egg-number–female-body-size function of the mountain populations was preserved in the mountain transplants; at their environmentally reduced sizes, mountain transplants produced the number of eggs expected from small *mountain* females, not lowland females of equivalent body size.

The point of this discussion is that the large number of eggs observed in mountain wood frog clutches were not what they appeared to be. They were greater than lowland clutches only as an indirect consequence of environmental induction and genetic variation for larger adult body size. The direct action of selection on the genetic variation associated with the number of eggs per body size was to favor smaller egg numbers. The interaction between egg number and egg size, and the evolutionary causes of the current character states are discussed elsewhere (Berven 1982a, b).

The dissection of environmental and genetic components of egg size was less complex. The large mountain females were observed to lay larger eggs than lowland females. But egg size depended on adult female body size, just as egg number did. Covariance analysis of the variation in egg size between the two populations, holding adult female body size as the covariate, revealed a significantly higher regression for mountain residents than the regression for lowland females (figure 1.10, bottom). Both regressions had the same slope. The transplantation experiment showed that individuals from both populations retained their original egg-size–

Environmental Component

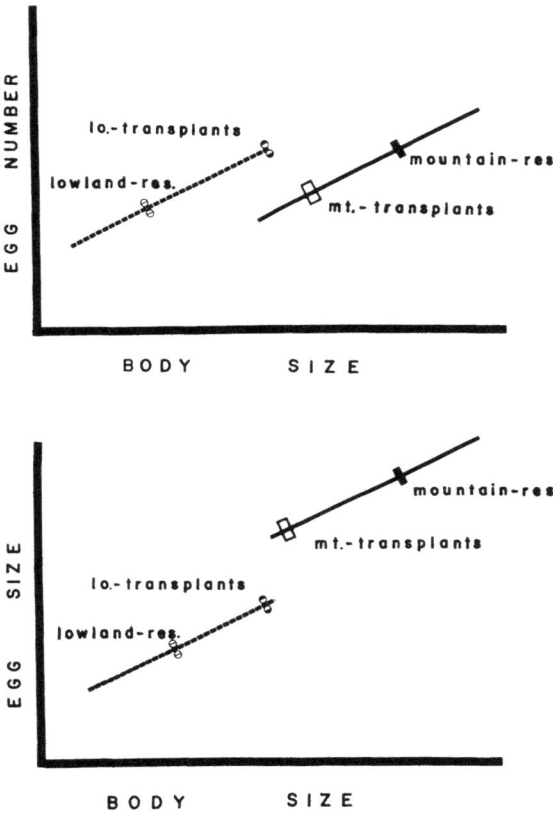

Figure 1.10. Comparison of the relationship of egg number to adult body size (above) and the relationship of egg size to adult body size (below) in two populations of the wood frog, *R. sylvatica*.

Note: In each graph the slopes of the mountain and lowland regressions are the same, but the adjusted means (heights) of the lines are significantly different. *Lowland-res.* = juveniles of lowland origin which developed in lowlands as lowland-resident controls. *Lo.-transplants* = juveniles of lowland origin transplanted to the mountains. *Mountain-res.* = juveniles of mountain origin which developed in mountains as mountain-resident controls. *Mt.-transplants* = juveniles of mountain origin transplanted to the lowlands. Means and ranges of each group are depicted. Figure is a facsimile of data from Berven (1982a).

body-size relationships despite the exaggerated and miniaturized body sizes brought about by environmental induction. Thus the large eggs of mountain wood frog females were the result of both environmental and co-gradient genetic variation.

To summarize these experiments, significant environmental influences were found in all but two of the life history traits examined. Characteristics of larval growth and development, the size and age of sexual maturity, and the size and number of eggs in a clutch were all strongly determined by the environment in which the individuals developed. However, genetic effects were also demonstrable (table 1.4). In the case of the green frog tadpole development, the genetic effects were counter to the observed gradient; in wood frogs they were co-gradient. The genetic component of all the life history traits of the wood frog proved to be co-gradient: wood frogs on tops of mountains are indeed genetically larger, slower at development, etc. It was particularly significant and unexpected that the pattern discovered in one species was not the same in a congener. Thus, the rules for the environment vs. genetic conflict were not even generalizable to species within a genus.

Conclusions

This article has presented data which illustrate that morphological, developmental, and life history traits are strongly influenced, if not determined, by the environment in which a developing or breeding individual finds itself. Only two out of ten life history traits examined had a strictly genetic basis. Specifically discussed were the breeding frequency and survivorship of red-spotted newts, the widespread reproductive failure of salamanders, frogs, and dragonflies, and the clines in developmental, morphological, and life history traits in frogs over an altitudinal gradient.

In the first case, the pattern of breeding frequency in female red-spotted newts was correlated with levels of infection by trypanosomes, a blood parasite. Level of infection is a characteristic acquired during ontogenesis. In the second case, reproduction at the populational level of a wide variety of aquatic species showed a boom-or-bust pattern, both among ponds and among years. In each case climatic and some biotic variables were implicated. From the point of view of the species (or individuals) them-

Environmental Component

Table 1.4. Comparison of Environmental and Genetic Variation in Developmental and Life History Traits in Two Species of *Rana* Frogs.

	Green Frog (*Rana clamitans*)			Wood Frog (*Rana sylvatica*)		
	Phenotype in Mountains Relative to Lowlands	Environmental Induction	Type of Genetic Variation	Phenotype in Mountains Relative to Lowlands	Environmental Induction	Type of Genetic Variation
1. Larval Size per Developmental Stage	larger	yes	mixed gradient minimizing	larger	yes	none
2. Larval Period	longer	yes	mixed gradient minimizing	longer	yes	none
3. Adult Size at Maturity						
Male	larger	n.t.	n.t.	larger	yes	Co-gradient
Female	larger	n.t.	n.t.	larger	yes	Co-gradient
4. Adult Age at Maturity						
Male	n.t.	n.t.	n.t.	older	yes	Co-gradient
Female	n.t.	n.t.	n.t.	older	yes	Co-gradient
5. Egg Number (Absolute)	n.t.	n.t.	n.t.	greater	yes	Co-gradient
Egg Number/♀ Body Size	n.t.	n.t.	n.t.	fewer	no	Co-gradient
6. Egg Size	n.t.	n.t.	n.t.	greater	yes	Co-gradient
Egg Size/♀ Body Size	n.t.	n.t.	n.t.	greater	no	Co-gradient

NOTE   n.t. stands for not tested.

selves, the variation was unpredictable in both space and time. The evolution of regulatory mechanisms may often be difficult or impossible.

In the last case, clinal variation in certain developmental and life history traits in frogs was analyzed for environmental and genetic components by laboratory experiments and manipulations of natural populations. In all but two of the traits studied, including larval growth rates and sizes at metamorphic climax, age and size of sexual maturity, and the size of eggs and clutches, strong environmental influences were responsible for much of the variation. Genetic determinants were significant, but in one species the direction of the genetic effects was consistent with the observed phenotypic cline and in a congener the genetic effects were antagonistic to it.

The implications of the observations and experiments presented here present serious challenges to the customary interpretations of "the adaptive significance" of traits seen in nature. Our finding strong environmental components and counter-gradient genetic variation in some life history characteristics should raise a yellow flag of caution to investigators who are too eager to assume the importance of evolution in the variation of life history trait patterns. It is not possible to gather comparative life history data from several populations by empirical methods alone, and correctly interpret the evolutionary basis of the observed variation. It is essential that the true genetic bases of the traits be dissected from the large quantities of variation induced by the ambient environment before the role of natural selection can be discussed. The effects of parasitisms must be eliminated. The occurrence of local, historical, catastrophes must be accounted for. Density dependence in fecundity and survivorship must be analyzed. Reciprocal transplantations of populations to the alternative environments, culture of populations in the laboratory under controlled conditions, and the manipulation of critical environmental variables are experimental procedures that are essential to separate character states which are inherited from states which are induced.

It is clear that character states themselves are usually the products of ambient environments, but perhaps the ability to be variable may be an evolved trait. The issues of ontogenetic plasticity and modulation versus tightly regulated, canalized development may be extremely important topics for evolutionary ecologists to consider in the future (Smith-Gill 1983).

ACKNOWLEDGMENTS

This work has been supported by grants from National Science Foundation, including BMS 74-19664, DEB 76-20326, DEB 77-04817, and DEB 78-10832.

A Sigma Xi grant to B. A. Mock, research funds from the Department of Zoology, and computer time from the Computer Science Center of the University of Maryland are gratefully acknowledged. Permission to study the populations in the George Washington National Forest has been kindly granted by District Ranger George H. Blomstrom. Permission to study the populations of frogs in the Beltsville Agricultural Station, Beltsville, Maryland has been possible through the assistance of Mr. Robert Almond. Field and laboratory assistance has been given by K. Brugger, D. Darling, R. Fritz, T. Halverson, R. Harris, C. Kitty, C. Molineaux, D. Norton, J. Sickel, C. Standbury, S. Thompson, G. Tong, W. Vanko, C. West, M. Wright, C. Zinter. The data on the dragonflies are attributable to T. Halverson. Conceptual as well as experimental inspiration was given by S. J. Smith-Gill. To all of these agencies and individuals we are greatly respectful and appreciative.

REFERENCES

Andrewartha, H. G. and L. C. Birch. 1954. *The Distribution and Abundance of Animals*. Chicago: University of Chicago Press.

Berven, K. A. 1981. Mate selection in the wood frog, *Rana sylvatica*. *Evolution* 35:707-722.

Berven, K. A. 1982a. The genetic basis of altitudinal variation in the wood frog, *Rana sylvatica*. I. An experimental analysis of life history traits. *Evolution* (in press).

Berven, K. A. 1982b. The genetic basis of altitudinal variation in the wood frog, *Rana sylvatica*. II. An experimental analysis of larval development. *Oecologia* (Berl.) 52:360–369.

Berven, K. A., D. E. Gill, and S. J. Smith-Gill. 1979. Counter gradient selection in the green frog, *Rana clamitans*. *Evolution* 33:609–623.

Birch, L. C. 1957. The role of weather in determining the distribution and abundance of animals. *Cold Spring Harbor Symp. Quant. Biol.* 22:203–218.

Brown, J. L. 1964. The evolution of diversity in avian territorial systems. *Wilson Bull.* 76:160–169.

Bryant, E. 1971. Life-history consequences of natural selection: Coles' result. *Am. Nat.* 105:75–76.

Charlesworth, B. 1971. Selection in density regulated populations. *Ecology* 52:469–474.

Charnov, E. L. and W. M. Schaffer. 1973. Life-history consequences of natural selection: Cole's result revisited. *Am. Nat.* 107:791–793.

Clausen, J., D. D. Keck and W. M. Hiesey. 1940. Experimental studies on the nature of species. I. The effect of varied environments on western North American plants. Carnegie Inst. Washington, Publ. No. 520.

Cole, L. C. 1954. The population consequences of life history phenomena. *Quant. Rev. Biol.* 29:103–137.

Emlen, S. T. and L. W. Oring. 1977. Ecology, sexual selection, and the evolution of mating systems. *Science* 197:215–223.

Fagen, R. M. 1972. An optimal life-history strategy in which reproductive effort decreases with age. *Am. Nat.* 106:258–261.

Gadgil, M. and W. H. Bossert. 1970. Life-history consequences of natural selection. *Am. Nat.* 104:1–24.

Gill, D. E. 1978a. The metapopulation dynamics of the red-spotted newt, *Notohthalmus viridescens* (Rafinesque). *Ecol. Monogr.* 48:145–166.

Gill, D. E. 1978b. Effective population size and interdemic migration rates in a metapopulation of the red-spotted newt, *Notophthalmus viridescens*, Rafinesque. *Evolution* 32:839–849.

Gill, D. E. 1978c. Occurrence of trypanosomiasis in the red eft stage of the red-spotted newt, *Notophthalmus viridescens* (Rafinesque). *J. Parasitol.* 64:930–931.

Gill, D. E. 1978d. On selection at high population density. *Ecology* 59:1289–1291.

Gill, D. E. 1979. Density dependence and homing behavior in the red-spotted newt, *Notophthalmus viridescens* (Rafinesque). *Ecology* 60:800–813.

Hairston, N. G., D. W. Tinkle, and H. M. Wilbur. 1970. Natural selection and the parameters of population growth. *J. Wildlife Management* 34:681-690.

King, C. E. and W. W. Anderson. 1971. Age-specific natural selection. II. The interaction between r and K during population growth. *Am. Nat.* 105:137-156.

Lack, D. 1954. *The Natural Regulation of Animal Numbers.* Oxford University Press, New York.

Lewontin, R. C. 1965. Selection for colonizing ability. In H. G. Baker and G. L. Stebbins, eds., *The Genetics of Colonizing Species*, pp. 77-94. New York: Academic Press.

MacArthur, R. H. 1976. *Geographic Ecology: Patterns in the Distribution of Species..* New York: Harper & Row.

MacArthur, R. H. and E. R. Pianka. 1966. On optimal use of a patchy environment. *Am. Nat.* 100:603-609.

MacArthur, R. H. and E. O. Wilson. 1967. *The Theory of Island Biogeography.* Monographs in Population Biology. Princeton, N. J.: Princeton University Press.

May, R. M. 1973. *Stability and Complexity in Model Ecosystems.* Monographs in Population Biology. Princeton, N. J.: Princeton University Press.

Maynard Smith, J. 1970. Genetic polymorphism in a varied environment. *Am. Nat.* 104:487-490.

Murphy, G. I. 1968. Pattern in life history and the environment. *Am. Nat.* 102:52-64.

Pianka, E. R. 1970. On r and K selection. *Am. Nat.* 104:592-597.

Pianka, E. R. 1972 r and K selection or b and d? *Am. Nat.* 106:581-588.

Pianka, E. R. and W. S. Parker. 1975. Age-specific reproductive tactics. *Am. Nat.* 109:453-464.

Roughgarden, J. 1971. Density dependent natural selection. *Ecology.* 52:453-468.

Roughgarden, J. 1974. Population dynamics in a spatially varying environment: How population size "tracks" spatial variation in carrying capacity, *Am. Nat.* 108:649-664.

Roughgarden, J. 1979. *Theory of Population Genetics and Evolutionary Ecology: An Introduction.* New York: MacMillan.

Schaffer, W. M. 1974. Selection for optimal life histories: The effects of age structure. *Ecology* 55:291-303.

Smith-Gill, S. J. 1983. Developmental plasticity: Developmental conversion versus phenotypic modulation. *Am. Zool.* (in press).

Stearns, S. C. 1976. Life history tactics: A review of the ideas. *Quarterly Review of Biology* 51:3-47.

Stearns, S. C. 1977. The evolution of life history traits: A critique of the theory and a review of the data. *Annual Rev. Ecol. Syst.* 8:145-171.

Wiens, J. A. 1976. Population responses to patchy environments. *Annual Rev. Ecol. Syst.* 7:81-120.

Wiens, J. A. 1977. On competition and variable environments. *Am. Scien.* 65(5):590-597.

Wilbur, H. M., D. W. Tinkle, and J. P. Collins. 1974. Environmental certainty, trophic level, and resource availability in life history evolution. *Am. Nat.* 108:805-817.

2 Clonal Diversity in Cladoceran Populations

PAUL D. N. HEBERT

Modern concepts in population biology are derived almost entirely from studies on sexually reproducing species. Asexual organisms offer the chance both to examine the adequacy of existing generalizations and to provide new insight into problems which are difficult to investigate in sexual organisms. Perhaps the merits and limitations of this approach can be shown most clearly by dealing with three problems—the role of heterosis in maintaining genetic variation, the limits to species richness, and finally the selective forces maintaining sexual reproduction.

There are three major taxa which reproduce by cyclical parthenogenesis—aphids, rotifers, and cladocerans. A few species in these groups have relinquished sex entirely and reproduce by obligate parthenogenesis. In addition there are a number of vertebrate and invertebrate taxa which contain species reproducing asexually. Genetic studies have been carried out on a fairly diverse assemblage of obligate parthenogens—including fish, weevils, moths, grasshoppers, etc.—but the only detailed work on cyclical parthenogens has dealt with cladocerans. Most of my conclusions on the genetics of cyclic parthenogens will, as a consequence, be based on these organisms, especially species of *Daphnia*. This genus includes some species which reproduce by cyclical parthenogenesis and others which are obligate parthenogens. The frequency of sexual reproduction in the cyclical parthenogens is determined by the nature of the environment they inhabit. Those living in permanent bodies of water reproduce sexually less often than those in intermittent habitats.

The Vexed Problem of Heterosis

The frequency of heterosis at variable gene loci is a matter of considerable importance in resolving the question of how genome variation is maintained. Three possibilities exist: heterosis may occur at nearly all variable loci, at a small percentage ($\approx 200-300$ loci), or at very few. Theoretical studies have shown that the first two possibilities are difficult to distinguish, for if the genome is studded with a few hundred heterotic loci, then associative overdominance will develop at loci with neutral alleles (Ohta and Kimura 1970, 1971).

Studies on sexually reproducing populations have revealed little evidence of heterosis (Lewontin 1974). Either heterosis exists at very few loci or the selection coefficients are too small to have detectable effects on genotypic frequencies. This latter possibility is very real. Lewontin (1974) pointed out that a sample size of more than 100,000 individuals of each genotype is needed to have a 95 percent chance of detecting a 1 percent heterozygote advantage. Assessing the prevalence of heterosis at variable loci in sexually reproducing populations is an intractable problem.

Populations which reproduce by cyclical parthenogenesis offer a much better chance of detecting small selection coefficients (Berger 1976). Table 2.1 shows that in the case where a heterozygote has a 1 percent fitness advantage, genotypic frequencies deviate markedly from Hardy-Weinberg expectations after as few as 50 generations of parthenogenesis. After several hundred generations, the population will consist nearly entirely of heterozygotes.

Allozyme studies have been carried out on populations of *Daphnia magna* inhabiting environments where continued parthenogenesis is possible. Five polymorphic loci have been studied

Table 2.1

Genotype	W	Genotypic Frequencies After X Generations of Asexual Reproduction			
		0	50	100	200
AA	0.99	0.25	0.187	0.134	0.059
Aa	1.00	0.50	0.625	0.732	0.882
aa	0.99	0.25	0.187	0.134	0.059

and at each locus heterozygote excesses were common (table 2.2). In fact significant excesses were noted in 28 and deficiencies in only 2 populations (Hebert 1972, 1974a,c; Hebert and Ward 1976; Young 1975, 1979a,b). While a surplus of heterozygotes is clearly the rule in these populations, Berger and Sutherland (1978) raised doubts about the generality of this conclusion based upon their own work with *Daphnia pulex* and other work on *Simocephalus serrulatus* (Smith and Fraser 1976). However, there is now strong evidence that many, if not all, populations of *D. pulex* in eastern Northern America reproduce by obligate parthenogenesis (Hebert and Crease 1980). The genetic data on *S. serrulatus* are also not compatible with the suggestion that this species reproduces by cyclical parthenogenesis (Hebert 1980; Hann and Hebert 1982).

Table 2.2. Hardy-Weinberg Deviations in Permanent Populations of *Daphnia magna*.

Locus	Number of Polymorphic Populations	Number of Populations with a Significant Heterozygote	
		Excess	Deficiency
ALK-2	1	1	
EST-1	15	9	1
GOT	4	4	
MDH	18	12	1
TO	3	2	
	41	28	2

Thus *D. magna* is the only cyclical parthenogen whose genetics have been investigated, and populations of this species inhabiting ponds where continued parthenogenesis is possible show clear evidence of heterozygote excesses at enzyme loci.

Angus (1978) has argued that, even if heterozygote excesses are documented at a few loci, it would be unwise to draw any conclusions of how selection acts on the genome as a whole. This argument assumes that selection operates on each locus in a unique way, as would be the case if selection were operating on fitness differences between genotypes at the locus under study. In clonal populations natural selection will undoubtedly remove alleles with major deleterious effects ($s \geqslant 0.20$) or fix alleles which lead to large increases in fitness, but the changes in frequency of alleles with small effects on fitness ($s \leqslant .01$) are likely to be largely stochastic. A mildly deleterious allele might, for instance, rise to high frequency because of linkage to favorable gene complexes. Genetic studies (Hebert 1975; Hebert and Moran 1980) have shown that because of founder effects, the allelic composition of each *Daphnia* population is unique. It follows that associations between genotypes will vary among populations, and thus a particular allele might rise to fixation in some populations, while being lost from others.

There is abundant evidence from studies on sexually reproducing species that most allozyme variants have small, if any, effects on fitness (Lewontin 1974). Allozyme genotypes in populations of *Daphnia* which regularly reproduce sexually also show no measurable fitness differences (Hebert 1974a; Hebert and Moran 1980), but in permanent populations, allozyme genotypes have been associated with selection coefficients of 0.30–0.50. Clearly these fitness differences are not caused by the allozyme variants themselves. If one accepts that allozyme variants at individual loci have inconsequential effects on fitness in relation to fitness differences generated via disequilibria, then a further conclusion follows. The study of a single locus can provide an indication of how selection operates in the genome where genotypes have small fitness differences. In short, the study of a malate dehydrogenase polymorphism in 10 different populations is likely to provide as much information about the action of selection as the study of a single different polymorphism in each of the populations. A parallel can be drawn to the study of inbreeding in sexually reproducing populations—in which the analysis of genotypic frequencies at a single gene locus where selection is unimportant is sufficient to provide information on the extent of inbreeding. In conclusion, the data favoring the general prevalence of heterozygotes is stronger than it seems. Heterozygote excesses are the rule at the five loci studied and this pattern is likely to hold for other loci.

A major question remains. Why should individuals heterozygous at an enzyme locus, which itself probably has little effect on fitness, be more fit than individuals homozygous at the locus? The simplest explanation is that the level of genome heterozygosity is important in determining fitness, and that a positive correlation exists between genome heterozygosity and heterozygosity at the enzyme locus. In short, heterozygosity at the enzyme locus is serving as a "marker" of enhanced heterozygosity in the genome. There are three possible causes of such a relationship. There can be little doubt that an individual heterozygous at a "marker" locus is more likely to be heterozygous at loci tightly linked to the marker than individuals homozygous for the marker. However, in this case one would expect the correlation with genome heterozygosity to be weak. Many clones heterozygous at the marker would have lower genomic heterozygosity than marker homozygotes. If genomic heterozygosity determined fitness, one would expect an excess of clones heterozygous at the marker, but their preponderance should not be great. A stronger correlation between heterozygosity at a given locus and genome heterozygosity can result in two ways. It would exist if heterozygosity at the marker locus resulted from a cross between two inbred lines with gene substitutions at many loci. Alternatively it could result if natural selection removed all but the most highly heterozygous clones generated via recombination (figure 2.1).

This assumes that there is a direct relationship between genomic heterozygosity and fitness. The increased fitness of highly heterozygous individuals produced by crossing inbred strains of normally outbred plant species is well documented. Similarly there is considerable evidence of heterosis in *Drosophila* species and to a

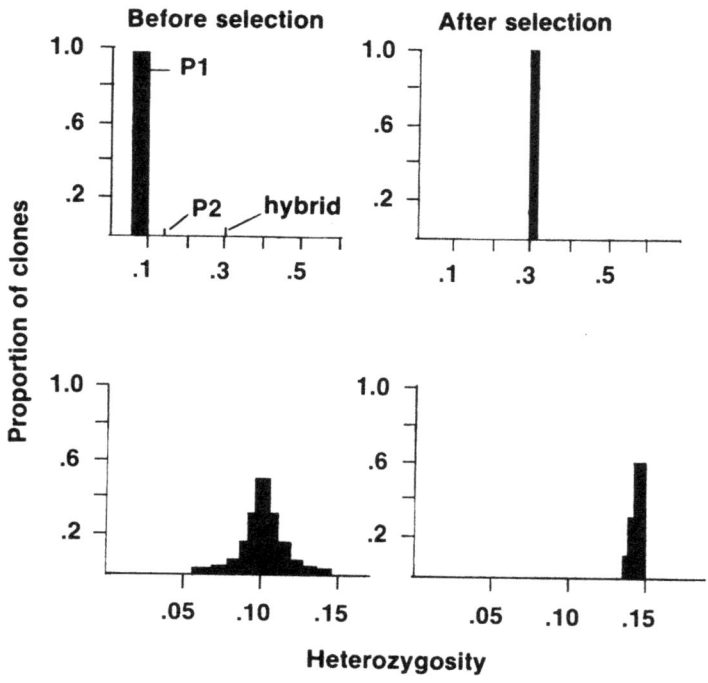

Figure 2.1. Two possible origins for the highly heterozygous genotypes found in *Daphnia magna* populations reproducing by continued parthenogenesis. In the first case heterozygotes originating from the hybridization of inbred lines are favored by selection. In the second case selection favors the most heterozygous genotypes produced by recombination.

certain extent in other normally outbred animal species. Lewontin (1958) suggested that heterosis did not occur in species which regularly inbreed, but the few studies carried out on inbreeding plants suggest that heterosis remains important (Jain and Allard 1960). There is strong evidence that *Daphnia* populations are inbred. Panmixis is approached within populations, but each includes only a small portion of the total species variability due to founder effect. The inbreeding coefficient due to population subdivision ranges from 0.2 to 0.4 in both *D. magna* and *D. carinata*. There have been no prior studies to determine if heterosis is exhibited by the offspring of crosses between genetically distinct lines of *Daphnia*. To investigate this matter, crosses were made between strains of *D. magna* from Canada and England. Genetic differences between the parent strains were marked. Genetic similarities based on study of 16 allozyme loci of English vs Canadian populations averaged only 0.6. The English clones had higher intrinsic rates of increase than the Canadian clones at the three test temperatures (10°, 20°, 30°C). The mean intrinsic rate of increase of the five hybrid clones exceeded that of their parents at all temperatures (figure 2.2). Additional experiments revealed that the hybrids survived environmental stresses more successfully than their parents (Hebert, Ferrari, and Crease 1982).

The hybridization studies indicate that offspring produced by crossing inbred lines of *Daphnia* show hybrid vigor. Natural populations are inbred, so it is possible that the heterozygote clones prevalent in nature derive from crosses between populations. However, there is no simple means of ruling out the possibility that the prevalence of heterozygote clones results from the selective elimination of clones with low heterozygosity. Distinguishing between these alternatives is significant: if the latter explanation is valid, then heterosis must exist at a minimum of several hundred loci, while if the former interpretation is correct, only a few loci need be heterotic. Regardless of its origin, there can be little doubt that heterosis is important in maintaining genetic variation in *Daphnia* populations. There is no reason to think that the way selection acts on *Daphnia* populations is in any way unusual. If large enough samples were analyzed, heterozygote excesses would also be the rule at variable loci in populations reproducing by obligate sexuality.

Why Are There So Few Species?

Ecologists and systematists have long been in awe of the vast number of different animal and plant species (Hutchinson 1959; Janzen 1970). This is understandable, for there are 3 or 4 orders of magnitude more species than the most competent taxonomist can identify. Nonetheless most species are represented by an enormous number of individuals. The modal number of individuals for North American bird species is approximately 10^5 (Preston

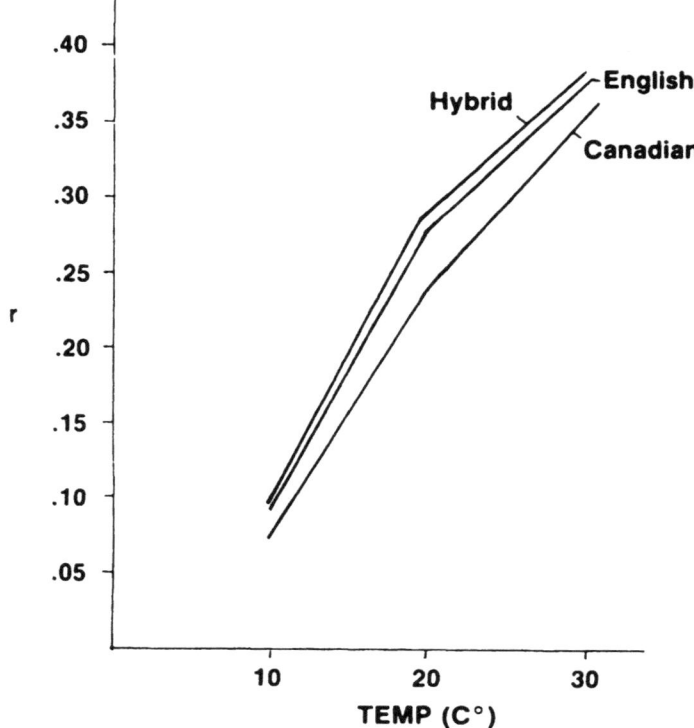

Figure 2.2. Intrinsic rates of increase for English, Canadian, and hybrid clones of *Daphnia magna*.

1948). If we assume that there are one thousand times as many macromoths as birds in an acre of forest, then the modal abundance of the North American species of these insects is likely to be about 10^7 (as there are about ten times as many moth as bird species). Such great abundances suggest that species richness has not been limited by the ultimate need for each species to have enough individuals to ensure local mating success and enough demes to cope with their stochastic extinction. If would seem that the biomass present on earth could be partitioned into many more species.

Competitive processes are often viewed as a potent means of reducing species diversity (MacArthur 1972). It has commonly been argued that one could not expect to find, for example, ten species of heron feeding on the same food items in the same locale, because one species would be better adapted and thus displace the others. If such competitive exclusion is frequent, then coexisting species should show niche separation. It is well established that coexisting bird species feed either on different food items or in different microhabitats (Cody 1974). Invertebrate communities provide similar instances of niche differentiation. Hutchinson (1951) argued that coexisting copepods are not random assemblages of available species, but that cohabiting species tend to differ markedly in size and that such size differences reduce food overlap. More recent workers (Miracle 1974; Lane 1975; Makarewicz and Likens 1975) have supported the notion that zooplankton communities include species which differ either in size or microhabitat preferences.

If competitive processes limit niche overlap, then clonal complexes should show distribution patterns like those of closely related sexual species. Certainly, it has been vigorously argued that only a single clone should occur in a particular habitat (Williams and Mitton 1973; Williams 1975). If clonal diversity exists, one would, on this basis, expect a checkerboard distribution pattern (Diamond 1975) with a single clone per habitat and the presence of any particular clone determined largely by chance.

It has recently been shown that at least some North American populations of *Daphnia pulex* reproduce by obligate parthenogenesis (Hebert and Crease 1980). In fact all eleven of the Ontario populations studied reproduced in this manner. Using electrophoresis, 22 clones were identified in these populations and as many as seven different clones were found in a single habitat. Coexisting clones were often genetically very distinct, indicating that clonal diversity was not the result of mutational events in each population. Populations of *D. middendorffiana* in Arctic Canada also usually include several different clones (Hebert, personal observation). Cladoceran populations are not exceptional—asexual "species" usually include many different clones, and these clones often coexist (Parker 1979; Mitter et al. 1979).

Such observations force a reassessment of the significance of the niche differentiation which normally exists between sexually reproducing species. The greater overlap evident in clonal organisms is likely to be a direct consequence of the ease of "speciation"— a single mutation suffices to produce a new reproductively isolated entity. By contrast, speciation in sexual organisms is a lengthy process, which itself often involves niche differentiation. Indeed sympatric speciation is a consequence of such differentiation. In the case of geographical speciation the population isolates often inhabit environments with different resources and (or) physical regimes and this is likely to promote ecological divergence. In summary conventional speciation fosters niche differentiation.

The data on clonal complexes suggest that competitive processes are of little importance in limiting species richness. Present-day species richness has undoubtedly been determined by the balance between the rate at which new species have been produced and the rate at which species have become extinct, but extinction rates have likely been independent of richness. Whittaker (1972, 1977) has similarly concluded from his study of species diversity patterns that plant and insect communities do not approach the richness at which competitive processes would become important in limiting diversity.

Obligate Parthenogens—Their Origins and Evolutionary Potential

A major impetus for work on parthenogenetic organisms has resulted from renewed interest in elucidating the selection pressures responsible for maintaining sexual reproduction. It has been argued for some time that populations which adopt obligate parthenogenesis are less able to adapt to environmental change than organisms reproducing sexually (Fisher 1930; Muller 1932, 1958; Crow and Kimura 1965). Support for the importance of such group selection comes from the taxonomic distribution of obligate parthenogens, for most have close relatives which reproduce sexually. Genera or higher categories which contain only asexual species are rare (Maynard Smith 1978) and this suggests that asexual taxa fail to persist long enough to adaptively radiate. The bdelloid rotifers are one striking exception to this generalization (Mayr 1963), but there may be other less spectacular examples. For example, males have not been found in the monogonont rotifer genus *Squatinella* (Pontin 1978) and all species in the oligochaete genus *Bimastos* (Jaenicke and Selander 1979) are asexual. Many similar examples might be given, but it can be argued that the occurence of sex in these groups has simply been overlooked. In defense it can be stated that there are also many taxa in which sex is presumed to occur without adequate evidence. Detailed genetic studies seem particularly essential to determine the frequencies of asexual taxa in the major groups of cyclic parthenogens.

The evolutionary potential of species which have adopted obligate parthenogenesis is an area of active study. Recent work has shown that asexual organisms are clonally diverse (Suomalainen and Saura 1973; Vepsäläinenan and Järvinen 1979; Vrijenhoek 1979; Mitter *et al.* 1979). However, in most cases it is difficult to determine whether this diversity evolved after the adoption of parthenogenesis or if it indicates the polyphyletic origins of asexuality. White (1966) has described cytological changes which likely occurred after the adoption of asexuality by the grasshopper *Moraba virgo*. It has been suggested that recessive lethals should accumulate in species reproducing parthenogenetically, but Suomalainen and Saura (1973) pointed out that their frequency cannot be determined because their detection requires the production of recombinants. There is, however, one class of recessive lethal whose presence can be detected electrophoretically. These are lethals at loci controlling the synthesis of central metabolic enzymes which possess a quaternary structure. In sexually reproducing populations null alleles are common at loci synthesizing peripheral enzymes, but have not been described at loci controlling central metabolic enzymes. Clearly individuals homozygous for this latter class of null alleles would die. Null alleles at loci controlling the synthesis of central metabolic enzymes do, however, exist in populations of *Daphnia* reproducing by obligate parthenogenesis. The best example involves the dimeric enzyme phosphoglucose isomerase. Heterozygotes are normally triple banded, but certain clones of *D. middendorffiana* show a two banded pattern (figure 2.3). This phenotype is most easily explained by assuming that these individuals are heterozygous for a normal

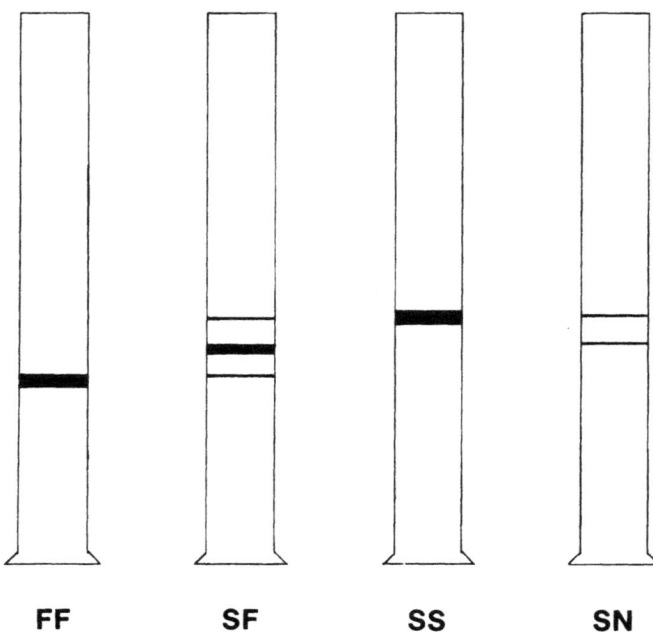

Figure 2.3. Phosphoglucose isomerase phenotypes in *Daphnia*.

allele and for an allele which produces a peptide chain with no enzyme activity. The association of null activity chain with a normal chain would result in the formation of a dimer with a single active site. If the nonfunctional chain had an altered charge, the mobility of this dimer would differ from that of dimers composed of two normal polypeptides. Each hybrid dimer would have half the catalytic activity of a normal dimer, but the hybrid molecules would be twice as common. Heterozygotes for such a null allele would be expected to have an electrophoretic phenotype with two equally staining bands—similar to the phenotype which is seen. The presence of null alleles at central metabolic loci provides good evidence of genome evolution subsequent to the adoption of parthenogenesis.

The rate of evolution is determined by the amount of genetic variation present and by the efficiency with which this variation is used. Mutation is the source of variation and there is good reason to expect higher mutation rates in asexual than sexual populations (Leigh 1970, 1973). Not only is variation likely to originate at higher rates in asexual populations, but it can also be utilized more efficiently. Under sexual reproduction only the additive component of the total variance in fitness is exploited, while under asexual reproduction the dominance and interaction components can be used as well. The opportunity which recombination provides to combine mutations that originate in different individuals is significant only when the population size is very large (Maynard Smith 1971; Kimura and Ohta 1971). Thus when environmental changes are gradual and population sizes moderate, there is little reason to expect that asexual organisms are at a disadvantage in relation to sexual species. If environmental changes occurred rapidly, it might seem that an asexual population would be doomed, while a sexual population could generate rare recombinants capable of survival. This conclusion overlooks the extent of genotypic diversity among clonal as opposed to sexual populations. Asexual species such as *D. pulex* include a relatively small number of genotypes, but the genetic distance among genotypes is great (figure 2.4). By contrast sexually reproducing populations of *Daphnia* contain a large number of different genotypes, but the divergence among genotypes is small. Atchley (1977) reports similar differences based on meristic and ecological studies of sexual versus asexual populations of grasshoppers in the genus *Warramaba*. The extended genotypic diversity of clonal organisms would seem to be a preadaptation to sudden environmental changes.

In sum, there seems to be little basis for concluding that asexual species have more limited abilities to adapt to environmental change than sexual species.

Group selection is not the only means by which sexual reproduction can be maintained. Indeed Williams (1975) has argued that sex can be maintained in cyclical parthenogens only if it offers some short-term advantage to balance the cost of meiosis. In the absence of such an advantage, Williams (1975) argues that clones which suppress meiosis should supplant their cyclical parthenogenetic ancestors.

Several points need to be considered before admitting the

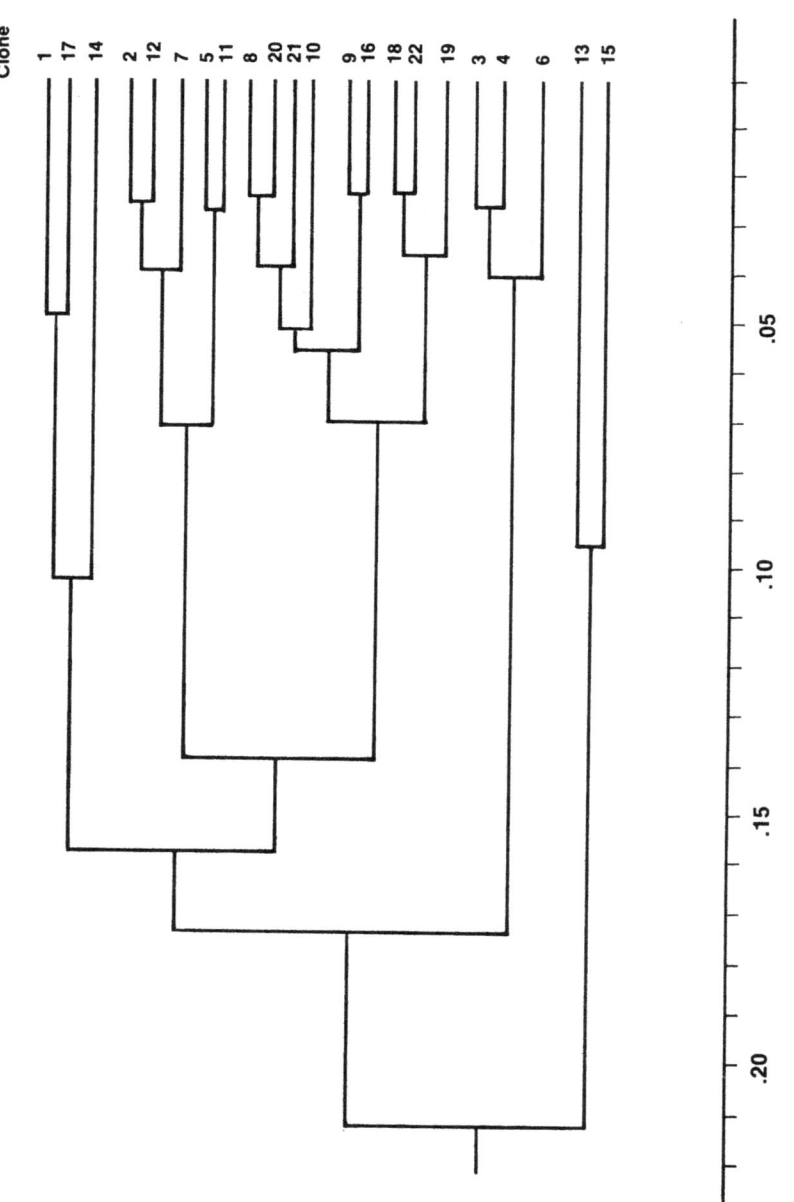

Figure 2.4. Single link cluster analysis of the genetic distances among clones of *Daphnia pulex*.

validity of this argument. Maynard Smith (1978) has pointed out that frequently there are important biological differences between the sexual and asexual eggs produced by a species. For instance, the sexual eggs of cladocerans are capable of diapause and are packaged in structures called ephippia, which resist desiccation, freezing, and digestive enzymes, while the asexual eggs develop immediately and are killed by exposure to harsh conditions. Maynard Smith (1978) stresses that if genetic variation is not present to decouple sex from such ecological correlates, it might be maintained as a consequence. In the genus *Daphnia* some species produce their ephippial eggs asexually and on this basis Maynard Smith concluded that the transition to asexuality in *Daphnia* should be frequent, unless sex conveys a benefit to balance the cost of meiosis. This conclusion was reached without considering either what it costs a clone to maintain sex or how sex might be lost.

Sex is less expensive for a cyclic parthenogen to maintain than for a sexual organism. Cyclic parthenogens pay the cost of meiosis only in the sexual generation and each bout of sex may be separated by numerous asexual generations. If 15 generations of parthenogenesis intervene between each sexual generation, then the cost of meiosis would be balanced by a 5 percent superiority in the mean fitness of sexual progeny. Secondly, the maintenance of sex need not halve the r_m of a clone, as it does in organisms with a 1:1 sex ratio. Sex in cladocerans is determined environmentally and males may constitute only a small proportion of the offspring produced by a female. Moreover male offspring are produced only as a population becomes crowded and there is reason to expect that female offspring produced under such conditions might be unable to reach maturity. Certainly adult males are much smaller than females, having only about 10 percent their biomass. The advantages to a cladoceran clone which dispenses with sexual reproduction therefore seem to be small.

However, as long as there is a finite cost in maintaining sex, it can be argued that the adoption of obligate parthenogenesis should be favored. In cladocerans, the transition to asexuality requires that two processes be terminated—meiosis during the production of ephippial eggs and male production in response to crowding. There is no doubt that these two processes are under separate genetic control and thus one process must be terminated before the other. A primary mutation suppressing male production must always be disadvantageous, regardless of the increase in ephippial egg production it causes, for males of other clones would fertilize the ephippia produced by such male-less clones, as well as any ephippia produced by females of their own clone(s). Thus mutations causing the loss of male production should not persist. A mutation suppressing meiosis would cause the production of both asexual ephippial eggs and sterile male offspring. Because of the retention of males, the asexual clone would not be expected to produce additional ephippia. In this case the gene terminating meiosis would not spread through the population, for the normal allele would replicate itself as frequently as the mutant allele. Indeed it would seem that mutations terminating male production and meiosis must occur synchronously if they are to spread. This is extremely unlikely, and thus it is fair to ask how obligate asexuality is ever able to evolve in cladocerans.

There is one particular class of mutations which can lead to a rapid spread of asexuality throughout local demes and indeed throughout the entire species population of a cladoceran. These are mutations which suppress meiosis during ephippial egg formation, but which fail to suppress sperm production by males. That is, the mutation terminating meiosis must be sex-limited in expression. Such mutations have an interesting property—the male carriers can spread the gene for asexuality onto a wide variety of genetic backgrounds.

The genetic consequences of the spread of such a mutation through a population can be readily simulated. Suppose we have a population which is polymorphic at a particular locus with two alleles, A^1 and A^2, at equal frequency. A second locus (M) controls meiosis—the normal allele (m) causes meiosis to occur, while a rare dominant mutation (M) suppresses meiosis in females. Further assume that the fitnesses of the three genotypes at locus A are identical, that genotypic frequencies at the locus are in Hardy-Weinberg equilibrium, and that the mutation suppressing meiosis does not affect fitness. A single mutation occurs at the M locus. This mutant will be the founder of a clone which produces ephippia asexually, but continues to produce fertile males. Random mating is assumed to occur between all male genotypes and the females which produce haploid ephippial eggs. When males from the asexual clone mate with such sexual females, half of the resulting offspring will be asexual. The increase in frequency of obligate asexual clones has been studied using a deterministic model. The results indicate that all individuals in the population would be obligate parthenogens after 46 generations if the population size was 10^5 (figure 2.5). Larger population sizes fail to materially slow the conversion to asexuality; with a population size of 10^6 individuals all are asexual after 55 generations. Once this conversion is complete, there are three clones in the population (A^1A^1, A^1A^2, A^2A^2) with frequencies that approximate Hardy-Weinberg equilibrium. The genotype at locus A of the first mutant individual has little effect on final clone frequencies. If a clone carrying the mutation were transferred to a second habitat, this population would be converted to asexuality in the same period of time. Clonal diversity would, of course, be likely to increase.

If meiosis is suppressed by a recessive mutation, the displacement of cyclical parthenogenetic clones requires more time. It would take 1920 generations for the mutation to reach fixation in a population of 10^5 individuals and 7902 generations in a population ten times as large. Again there would be three clones in the final population, with clone frequencies in Hardy-Weinberg equilibrium. In the case of a recessive mutation, the spread of asexuality would be much more rapid in secondary colonized habitats, because the colonist would be homozygous for the mutation. In this case it would take 265 generations to convert a population of 10^5 individuals to asexuality and 544 generations if the population included 10^6 individuals.

There is a fair amount of evidence which supports the credibility of this model. Firstly *Daphnia* clones which produce asexual ephippia often retain the ability to produce males (Hebert 1981a). The suppression of meiosis was not preceded or even accompanied by the loss of males. It remains uncertain whether these males are able to produce viable sperm. Genetic studies on *Daphnia* species reproducing by obligate parthenogenesis also seem compatible with this model. Clonal diversity is high and gene substitution is

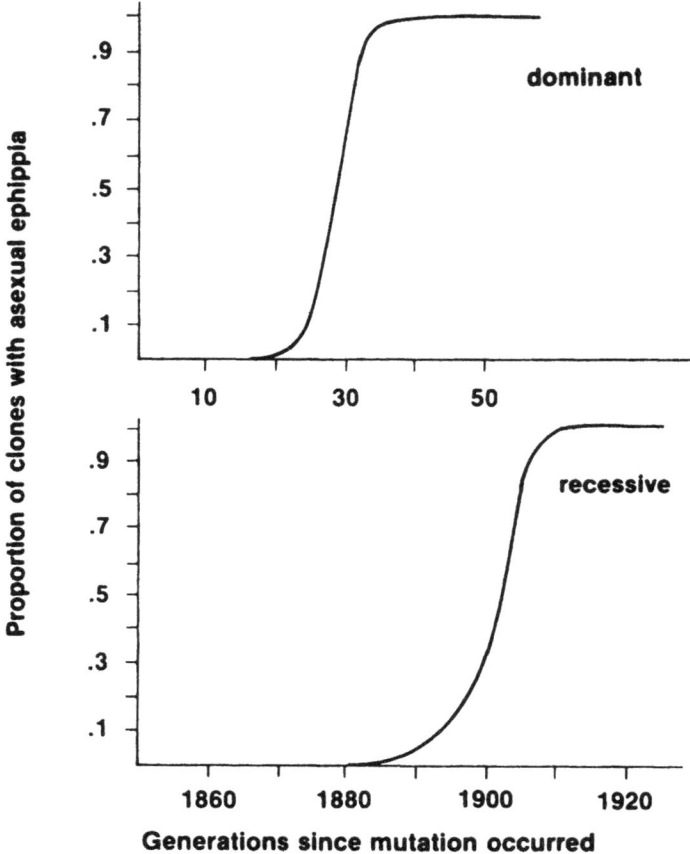

Figure 2.5. The increase in frequency of a mutation which suppresses meiosis only in females. The population size is 10^5.

prevalent between clones (Hebert and Crease 1980). The evidence against a monophyletic origin of these clonal complexes is compelling. However, it is admittedly difficult to distinguish between the polyphyletic origin of obligate parthenogenesis and the contagious spread of a single mutation conferring asexuality.

The means by which sexuality has probably been lost in the cladocerans is unlikely to be unique. Male production and sexual egg production are also distinct processes in the aphids, and thus asexuality might spread in a similar fashion in this group. The model does not apply to rotifers or cynipid wasps for in these taxa males are haploid and the termination of meiosis also stops male production. However, there are other taxa which may have adopted asexuality in a similar manner. Richards (1973) proposed a similar model to account for the spread of asexuality in the plant genus *Taraxacum* and his model may be broadly applicable to monoecious plants. Jaenicke and Selander (1979) use a related model to explain the loss of sexual reproduction in oligochaete worms and their model may apply to many hermaphroditic species. In all these cases the avoidance of the cost of meiosis is a consequence rather than a cause of the loss of sexuality. If the loss of sex is unrelated to the cost of meiosis, it follows that the maintenance of sex need not be related to any short-term advantage conferred by sex. There is, for example, in the case of cladocerans, no simple means of avoiding the cost of meiosis.

In the comparison of their evolutionary potential, it was pointed out that clonal morphospecies are often genetically more diverse than sexually reproducing species. When genotype–environment interactions in fitness are pronounced, clonal complexes should be at an advantage, but only so long as each habitat is colonized by a significant proportion of the different clones. In the case of cladocerans the evidence suggests that most habitats are colonized by few propagules. The few clones reaching a habitat will often have low fitness. A purely asexual clone's adaptation via mutation will be slow compared with that of a clone capable of sexual reproduction which can produce high fitness genotypes via recombination. The polyclonal nature of populations reproducing by obligate parthenogenesis indicates a second disadvantage of asexuality— uniclonal populations are inefficient in utilizing niche space. Efficient occupation of niche space must await the arrival of the appropriate secondary colonists. By contrast a clone capable of sexual reproduction can generate an array of daughter clones with differing ecological characteristics, and occupy the niche space in a short period of time. These arguments suggest that a clone which retains the ability to reproduce sexually will be at an advantage when vagility is limited.

The arguments in this section indicate that a number of factors are important in explaining the incidence of cladoceran species reproducing by obligate parthenogenesis. It has been shown that the loss of sexuality is likely to be caused only by a sex-limited mutation suppressing meiosis. The frequency of such mutations will clearly have an important bearing on the frequency with which asexuality is adopted. The relative persistence of species capable of sexual reproduction versus those which are obligate parthenogens is a second factor of importance. It seems likely that in some circumstances asexual populations may have a better chance of survival than populations capable of sex. For instance, because of their increased genotypic diversity clonal complexes probably stand a better chance of surviving an abrupt, widespread change in environmental conditions. Similarly, as long as their vagility is high, obligate parthenogens may also be at an advantage when genotype-environment interactions in fitness are great. However, if vagility is low, clones capable of sex will have a clear advantage. While certain factors have now been identified as being important in determining the frequency of asexual forms, an assessment of their relative importance remains an area for future work. The completion of these studies should provide a good understanding of the evolutionary forces responsible for the distribution of sexual forms in the Cladocera.

REFERENCES

Angus, R. A. 1978. *Daphnia* and the search for heterosis. *Am. Nat.* 112:955-956.

Atchley, W. R. 1977. Evolutionary consequences of parthenogenesis: Evidence from the *Warramaba virgo* complex. *Proc. Nat. Acad. Sci. U.S.A.* 74:1130-1134.

Berger, E. 1976. Heterosis and the maintenance of enzyme polymorphism. *Am. Nat.* 110:823-839.

Berger, E. and J. Sutherland. 1978. Allozyme variation in two natural populations of *Daphnia pulex. Heredity* 41:13-23.

Cody, M. L. 1974. Competition and the structure of bird communities. Princeton, N.J.: Princeton University Press.

Crow, J. F. and M. Kimura. 1965. Evolution in sexual and asexual populations. *Am. Nat.* 99:439-450.

Diamond, J. M. 1975. Assembly of species communities. In M. L. Cody and J. M. Diamond, eds., *Ecology and Evolution of Communities*. Cambridge, Mass.: Belknap Press.

Fisher, R. A. 1930. *The Genetical Theory of Natural Selection*. Oxford: Clarendon Press.

Hann, B. J. and P. D. N. Hebert. 1982. Reinterpretation of genetic variation in *Simocephalus* (Cladocera, Daphniidae). *Genetics*, in press.

Hebert, P. D. N. 1972. Enzyme Variability in Natural Populations of *Daphnia magna*. Ph.D. dissertation, Cambridge University.

Hebert, P. D. N. 1974a. Enzyme Variability in Natural Populations of *Daphna magna*. II. Genotypic frequencies in permanent populations. *Genetics* 77:323–334.

Hebert, P. D. N. 1974b. Enzyme Variation in Natural Populations of *Daphna magna*. III. Genotypic frequencies in intermittent populations. *Genetics*. 77:335–341.

Hebert, P. D. N. 1974c. Ecological differences between genotypes in a population of *Daphnia magna*. *Heredity* 33:327–337.

Hebert, P. D. N. 1975. Enzyme Variability in Natural Populations of *Daphnia magna*. I. Population structure in East Anglia. *Evolution* 28:546–556.

Hebert, P. D. N. 1980. The genetics of cladocera. In W. C. Kerfoot, ed. *The Evolution and Ecology of Zooplankton Communities*. Hanover, N.H.: University Press of New England.

Hebert, P. D. N. 1981. Obligate asexuality in *Daphnia*—is it contagious? *Am. Nat.* 117:784–789.

Hebert, P. D. N. and T. C. Crease. 1980. Clonal coexistence in *Daphnia pulex*—another planktonic paradox. *Science* 207:1363–1365.

Hebert, P. D. N. and C. Moran. 1980. Enzyme variability in natural populations of *Daphnia carinata* King. *Heredity* 45:313–321.

Hebert, P. D. N., C. Ferrari, and T. J. Crease. 1982. Heterosis in Daphnia: a reassessment. *Am. Nat.* 119:427–434.

Hebert, P. D. N. and R. D. Ward. 1976. Enzyme variability in natural populations of *Daphnia magna*. IV. Ecological differentiation and frequency changes of genotypes at Audley End. *Heredity* 36:331–341.

Hutchinson, G. E. 1951. Copepodology for the ornithologist. *Ecology* 32:571–577.

Hutchinson, G. E. 1959. Homage to Santa Rosalia or why are there so many kinds of animals? *Am. Nat.* 93:145–159.

Jaenicke, J. and R. K. Selander. 1979. Evolution and ecology of parthenogenesis in earthworms. *Am. Zool.* 19:729–737.

Jain, S. K. and R. W. Allard. 1960. Population studies in predominantly self-pollinated species. I. Evidence of heterozygote advantage in a closed population of barley. *Proc. Nat. Acad. Sci. U.S.A.* 46:1371–1377.

Janzen, D. H. 1970. Herbivores and the number of tree species in tropical forests. *Am. Nat.* 104:501-528.

Kimura, M. and T. Ohta. 1971. *Theoretical Aspects of Population Genetics*. Princeton: Princeton University Press.

Lane, P. A. 1975. The dynamics of aquatic systems: A comparative study of the structure of four zooplankton communities. *Ecol. Monogr.* 45:307-336.

Leigh, E. G. 1970. Natural selection and mutability. *Am. Nat.* 104:301-305.

Leigh, E. G. 1973. The evolution of mutation rates. *Genetics*, Suppl. 73:1-18.

Lewontin, R. C. 1958. Studies on heterozygosity and homeostasis II. Loss of heterosis in a constant environment. *Evolution* 12:494-503.

Lewontin, R. C. 1974. *The Genetic Basis of Evolutionary Change*. New York: Columbia University Press.

MacArthur, R. H. 1972. *Geographical Ecology*. New York: Harper and Row.

Makarewicz, J. C. and G. E. Likens. 1975. Niche analysis of a zooplankton community. *Science* 190:1000-1003.

Maynard Smith, J. 1971. What use is sex? *J. Theoret. Biol.* 30:319-335.

Maynard Smith, J. 1978. *The Evolution of Sex*. Cambridge University Press.

Mayr, E. 1963. *Animal Species and Evolution*. Cambridge, Mass.: Belknap Press.

Miracle, M. R. 1974. Niche structure in freshwater zooplankton: A principal components approach. *Ecology* 55:1306-1316.

Mitter, C., D. J. Futuyma, J. C. Schneider and J. D. Hare. 1979. Genetic variation and host plant relations in a parthenogenetic moth. *Evolution* 33:777-790.

Muller, H. J. 1932. Some genetic aspects of sex. *Am. Nat.* 68:118-138.

Muller, H. J. 1958. Evolution by mutation. *Bull. Am. Math. Soc.* 64:137-160.

Ohta, T. and M. Kimura. 1970. Development of associative overdominance through linkage disequilibrium in finite populations. *Genet. Res.* 16:165-177.

Ohta, T. and M. Kimura. 1971. Behaviour of neutral mutants influenced by associated overdominant loci in finite populations. *Genetics* 69:247-260.

Parker, E. D. 1979. Ecological implications of clonal diversity in parthenogenetic morphospecies. *Am. Zool.* 19:753-762.

Pontin, R. M. 1978. Key to the freshwater planktonic and semi-planktonic

rotifers of the British Isles. Freshwater Biological Association Publication 38.
Preston, F. W. 1948. The commonness and rarity of species. *Ecology* 29:254-283.
Richards, A. J. 1973. The origin of the *Taraxacum* agamospecies. *Bot. J. Linn. Soc.* 66:189-211.
Smith, M. Y. and A. Fraser. 1976. Polymorphism in a cyclic parthenogenetic species: *Simocephalus serrulatus*. *Genetics* 84:631-637.
Suomalainen, E. and A. Saura. 1973. Genetic polymorphism and evolution in parthenogenetic animals. I. Polyploid Curculionidae. *Genetics* 74:489-508.
Vepsäläinen, K. and O. Järvinen. 1979. Apomictic parthenogenesis and the pattern of the environment. *Am. Zool.* 19:739-751.
Vrijenhoek, R. C. 1979. Factors affecting clonal diversity and coexistence. *Am. Zool.* 19:787-797.
White, M. J. D. 1966. Further studies on the cytology and distribution of the Australian parthenogenetic grasshopper, *Moraba virgo*. *Revue Suisse Zool.* 73:383-398.
Whittaker, R. H. 1972. Evolution and measurement of species diversity. *Taxon* 21:213-251.
Whittaker, R. H. 1977. Evolution of species diversity in land communities. *Evolutionary Biology* 10:1-67.
Williams, G. C. 1975. Sex and Evolution. Princeton: Princeton University Press.
Williams, G. C. and J. B. Mitton. 1973. Why reproduce sexually? *J. Theoret. Biol.* 39:545-554.
Young, P. W. 1975. Enzyme polymorphism and reproduction in *Daphnia magna*. Ph.D. thesis, University of Cambridge.
Young, J. P. W. 1979a. Enzyme Polymorphism and Cyclic Parthenogenesis in *Daphnia magna*. I. Selection and clonal diversity. *Genetics* 92:953-970.
Young, J. P. W. 1979b. Heterosis following sexual reproduction. *Genetics* 92:971-982.

3 The Extent and Consequences of Heritable Variation for Fitness Characters

CONRAD A. ISTOCK

Two kinds of genetic analysis have coexisted within evolutionary biology since the beginning of its systematic development in the nineteenth century. These types of analysis represent the Mendelian genetic and quantitative genetic threads of modern population genetics, and both spring from the insights of Wallace, Darwin, and Mendel. Mendel immediately associated his empirical and theoretical discoveries with evolution. He saw his studies as bearing on "a question of the importance of which cannot be overestimated in connection with the history of the evolution of organic forms" (Mendel 1865; fourth paragraph, from a translation by William Bateson reprinted in Sinnot, Dunn, and Dobzhansky 1958). Near the end of his pea paper, when discussing the genetic divergence of "species" (varieties), Mendel is clearly again using the ideas argued by Wallace and Darwin only a few years earlier. This interrelation of evolutionary thought and genetic analysis remains as strong today as it was in Mendel's time, with added dimensions now coming from modern molecular genetics.

Beside the Mendelian mode of genetic analysis stands the independently developed quantitative genetics of Galton and the biometricians of the late nineteenth century. This approach to the study of the inheritance of metric or continuously varying characters clearly arose directly from Darwinism, and led to the notorious dispute between the biometricians and Mendelians in the early twentieth century (Provine 1971). Nilsson-Ehle (1909), East (1910), and Fisher (1918, 1958) resolved the dispute by recognizing that the additive influences of independently segregating Mendelian alleles at many loci could result in a continuous distribution of phenotypes for the morphological, physiological, behavioral, and ecological attributes of plants and animals. We now refer to this aggregative form of genetic transmission as quantitative or polygenic inheritance. Thus the implicit definition of a polygene is a multilocus set where each allele at each locus carries an incremental contribution to the mean and variance of the phenotypic trait in question.

One of the most striking and intriguing facts to emerge from studies in quantitative genetics during the past three decades is the wide extent of polygenic variability expressed for fitness characters in laboratory and domesticated populations, and possibly in natural populations as well. The primary fitness characters are survival probabilities, development times, and fertilities associated with particular genotypes in specific environments. Together the fitness characters describe both patterns of reproduction and the potential for natural selection. The plasticity of metric characters is at once the dilemma and the strength of quantitative genetic studies. Unlike discretely Mendelizing characters which typically, though not always, seem inflexible in expression over wide environmental ranges, metric characters are thought to be exceptionally variable in their heritabilities in different or changing environments (Feldman and Lewontin 1976). However, environmentally induced shifts in the phenotypic variance, genetic variance, and heritability of a trait provide a natural assessment of the complete interaction of genes and environments.

Parsons (1974) forcefully argued that male mating speed may be the most important component of fitness in *Drosophila spp.*, and that mating speed is always subject to directional selection for

Heritable Variation

rapidity. Parsons' contention is unarguable under *certeris paribus* conditions which must ensure that faster mating will bring more matings and more progeny per male lifetime. The relation between male mating speed and the primary fitness characters, however, is not obvious. In studies with *D. robusta*, Prakash (1967) found a significant positive correlation between male mating speed and fertility. If a positive association between male mating speed and all three primary fitness characters exists, Parsons' contention seems correct. Manning (1963) found that it was possible to select for slower, but not for faster, mating speed. This result at first seems to suggest past directional selection for fast mating, but one wonders why the residual variation for mating speed exists in sufficient quantity to support selection for slow mating speed. It remains thinkable that natural populations of *Drosophila spp.* and other organisms with a scramble-like mating process have been subject to stabilizing rather than directional selection. The role of female influences upon selection for male mating speed needs to be considered.

Many traits other than the primary fitness characters have been of interest in quantitative genetic studies. For practical reasons in animal husbandry, as well as for reasons from ecological theory involving foraging efficiency, prey defense, interspecific competition, and niche differentiation, the polygenic inheritance of body size has assumed a good deal of importance (cf. Robertson 1954; Kearsey and Kojima 1967; Nordskog 1977; Boag and Grant 1978). However, at this stage in our understanding we must be careful not to generalize about microevolution from the quantitative genetic variation for the morphological or other superficial, or peripheral (Robertson 1955), features of organisms frequently analyzed by quantitative geneticists in the past. Such variation, even if strongly heritable in the prevailing environment, cannot be caught up in microevolutionary change by natural selection if it has no covariance with heritable variation for one of the primary fitness characters. Here is a strict sense in which heritable variation, for traits other than fitness characters, may be neutral under natural selection (O'Donald 1971). It thus becomes necessary in all cases to demonstrate the connection of superficial traits and fitness or fitness characters as Kearsey and Barnes (1970) and Linney et al. (1971) have done for sternopleural chaeta number in *Drosophila melanogaster*. One might guess that many of the enzyme variants detected by electrophoresis and related methods are actually uncoupled from fitness and selection for substantial periods of evolutionary time, though we surely don't know this yet. Under artificial selection any genetically variable trait, Mendelian or polygenic, can be made into a surrogate for fitness. Artificial and natural selection probably differ in average intensity (Bulmer 1971), but they differ more fundamentally because the natural physiological, ontogenetic, and ecological relations between superficial characters and primary fitness characters are distorted under artificial selection. One might view artificial selection as a human game played, often to great benefit, with hopeful monsters.

In the study of natural selection an important distinction can be made between the "actual" and "potential" polygenic variation of a population. The actual variation is that which is currently expressed in the prevailing environment, while potential variation provides a reservoir of polygenic variation that will be manifest only after another round of recombination or following environmental change. Lande (1976) has made a distinction between "expressed" (i.e., actual) and "hidden" polygenic variability under stabilizing selection. Hidden variability is due to genetic expression which may occur in the prevailing environment, but is temporarily masked because combinations of closely linked alleles are negatively correlated for their effects on the phenotype. This cancellation effect in genetic expression arises because recombination continuously reshuffles the alleles favored by stabilizing selection for an intermediate optimum. I have elsewhere presented evidence that such stabilizing selection for variation in development time and diapause does occur in natural populations of the pitcher-plant mosquito, *Wyeomyia smithii*, and further that the capacity of a population to undergo continuous microevolutionary revisions of its intermediate optima for development time and diapause through recombination and stabilizing selection depends on individual genetic (mixed progeny) adaptation to a fluctuating environment (Istock et al. 1976a and b; Istock 1978, 1980). We

Heritable Variation

will use the term "potential" variation to encompass "hidden" variation in Lande's sense, as well as the additional variation released by altered effects on gene expression after environmental change. In future studies it will be necessary to separate these two forms of potential variation because the hidden form may continuously recur in the prevailing environment, while the form of potential variation released only by environmental change may be a central part of the genetic basis for longer term evolutionary adaptation. If this conjecture about two types of potential variation is correct, natural populations may store large reserves of variation, in the polygenic form, which is manifest only with environmental change. At this point, such speculations serve mostly to emphasize our need to know much more about the nature of polygenic variation.

The importance of an understanding of the genetic limits to fitness character variation for all of the theory of life history evolution was forcefully put by Stearns (1976). "There is a world of difference between the optimal and the best available. We can be sure that selection will elicit a local optimum from the variability available in the ancestral population, but we cannot be sure that the result will match predictions made by a mathematical model that does not take the limits of natural variability into account."

Let me draw together the main conclusions and suggestions thus far. Despite the difficulty of quantitative genetic analyses, and the abstractness of gene expression embodied in such thinking, (1) the environmental lability of quantitative genetic inheritance, (2) the direct and delayed interrelations between selection and polygenic systems, and (3) the central position of fitness characters in any model of microevolution taken together offer ecological and genetic realism. Here there may be the possibility of a synthesis, which has not been obtainable with strictly Mendelian approaches to problems of evolutionary genetics.

Some of the questions we want to answer are the following.

1. How do the genes behave statistically in the control of expression for metric fitness characters?

2. How are superficial metric traits, and traits controlled by simple one or two locus Mendelian inheritance, coupled with the

fitness characters through the entire system of genetic control and variation? Conversely, to what extent are variability for superficial and fitness traits uncoupled?

3. How much genetic variability for polygenic traits, and for fitness characters in particular, is held in natural populations?

4. How much polygenic variability in each generation is of the "actual" sort and how much is of the "potential" sort, and how much is "hidden" in Lande's (1976) sense?

5. How do stabilizing selection, mutation and recombination shape the patterns of fitness character variation found in natural populations?

6. What happens to polygenic systems when some fitness characters are subject to directional selection, as survival rates seemingly should be all the time (Mather 1973), while at the same time other fitness characters are subject to stabilizing selection?

7. How do the amounts and limits of fitness character variation influence life-history evolution?

The purpose of this paper is to assess past progress and future prospects for each of these questions. I will include only studies with animal populations. However, I do not believe that progress employing plant material is any more advanced than the animal work.

What Do We Suppose the Genes Do?

In quantitative genetics, abstract forms of gene expression and gene interaction are assumed. Each Mendelian allele is imagined to make additive, dominance, and epistatic contributions to the final phenotypic values of individuals as well as to the distribution of phenotypic values for the population. These contributions may be positive, negative, or zero. The theoretical magnitudes of the allelic contributions to a character are imagined to be scaled as a function of the number of alleles at a given locus and their respective frequencies. Furthermore, the magnitude of the contribution from one allele may be epistatically conditioned by contributions from alleles at other loci. Although there is no real understanding of gene expression involved here (Slatkin 1970), quantitative genetic methods have allowed the effective description, and manipulation of polygenic inheritance.

The current knowledge of gene expression in molecular and developmental genetics is, at present, inadequate to bridge the gap between molecular and quantitative genetics. We are not yet even aware of the scale or intricacy of the problems we face in trying to remove this gap. One wonders if a statistical description of the genetics of complex traits will still be required once the gap is closed. If the description remains statistical, the abstract and statistically tidy gene expression of quantitative genetics may not suffice to portray the relation of phenotypes and polygenic inheritance. At the moment, however, it is difficult to construct a solid theoretical connection between polygenic inheritance and knowledge of gene expression from molecular genetics, though Lewontin (1977) has argued that such knowledge should be of continuing relevance to quantitative geneticists. Above the molecular genetic level, some Mendelian loci with marker genes can be shown to have specific effects on quantitative traits (Thoday 1977). Such loci have been called "quantitative trait loci" by Gelderman (1975). At a still higher level of genetic organization there must be connections between the polygenes and chromosomes. Our knowledge of this connection, though still rudimentary, has been increased by a number of fascinating studies over the last two decades, and many of these studies have focused on the mapping of polygenic variation for fitness characters to the chromosomes. The justification for the "black box" approach of quantitative genetics may not only be the small amount of genetic reasoning it requires for the creation of coherent representations of polygenic inheritance (Lewontin 1977), but even more importantly its capacity to construct the kinds of statistical representations which we think are akin to those involved in natural selection and evolutionary change. After all, we think that natural selection is a process which continuously reviews the variation in fitness between phenotypes within the context of the entire population of phenotypes.

The central quantity in Fisher's (1918, 1958) theoretical development of the notion of genetic variance for a metric character is the deviation from the mean character value, or average effect, due to expression of each allele. Call this deviation $a(i)$ so that:

$$a(i) = [x(i) - \bar{x}]$$

where $x(i)$ is the mean phenotypic value associated with allele i over all genetic backgrounds and $\bar{x}$ is the mean phenotypic value for the whole population. Now rescale all phenotypes so $\bar{x} = 0$ then:

$$a(i) = x(i),$$

and the additive genetic variance $V(a)$ over all loci is:

$$V(a) = \sum_i p(i)a(i)^2 = \sum_i p(i)x(i)^2,$$

with $p(i)$ the frequency of the ith allele at its own locus. If the character of interest is fitness or one of its components, it is this variance which declines to zero as the mean fitness of a population smoothly rises according to Fisher's fundamental theorem of natural selection (Fisher 1958; Crow and Kimura 1970).

Now suppose that selection for an intermediate optimum always moves $x(i)$ and therefore $a(i)$ to zero for all the fitness characters (Wright 1968:272, "The character upon which selection operates directly, overall selective value, almost certainly depends on deviation from an intermediate optimum in nature"). There will be no additive variance for fitness in Fisher's sense. Does this mean there is no genetic variation for fitness at all?

Let $a(i, j)$ represent the average effect of the ith allele upon some fitness character when the ith allele is in the jth genetic background. Then:

$$a(i, j) = [x(i, j) - \bar{x}] = x(i, j), \bar{x} = 0,$$

with $x(i, j)$, the average phenotypic score of individuals with the ith allele in the jth genetic background, creating positive or negative genetic deviations in fitness as a consequence of recombination. Then, there may persist an additive genetic variance in fitness which cannot sustain directional selection over the generations, but which nevertheless repeatedly exposes some hidden variability in fitness each generation. As Lande (1976, 1977) and Maynard Smith (1979) have argued, mutation may be sufficient to maintain this kind of polygenic variation. A different genetic variance is measured by:

$$\acute{V}(a) = \sum_j \sum_i p(i,j) a(i,j)^2.$$

This $\acute{V}(a)$ does not drive fitness ever higher, but it may seek a persistent equilibrium value in Lande's sense, and it may be transformed to $V(a) > 0$ in Fisher's sense when a directional change of environment ensues. $\acute{V}(a)$ is not simply genetic load. In a temporally fluctuating environment it allows continuous genetic adjustment to environmental uncertainty (Istock 1978, 1980).

Populations in frequently or continuously fluctuating environments may actually accumulate additive genetic variance of the $\acute{V}(a)$ form as selection favors improved approximation of fluctuating intermediate optima over successive generations. Not only superficial characters but even fitness characters, such as developmental time and fertility, may seek such intermediate optima. Lethals and strongly deleterious alleles will still be eliminated, but alleles with small $x(i, j)$ will be accumulated, and potential variation may increase. This pattern of selection is one of fluctuating-stabilizing selection. Earlier designations such as quasi-normal, mildly deleterious, supervital, and subvital (Dobzhansky 1970; Lewontin 1974) to describe the fitness effects of variant alleles or chromosomes in a single, constant environment lose their meaning if variation and selection assume this latter configuration.

Studies of Variation in Fitness Characters

Those familiar with empirical studies in quantitative genetics know that most, if not all, of the past work has explored polygenic variation in laboratory environments or under conditions of agriculture or animal husbandry. Even nonapplied research in quantitative genetics has been done almost entirely in environments quite unlike those in which natural populations of the same species exist (c.f. Pollak, Kempthorne, and Bailey 1977). Thus we may have observed the release of potential polygenic variation for the most part. An important type of study for the future, particularly for the analysis of fitness characters, will depend on measurement of actual polygenic variation measurements made directly in natural populations, or at least in carefully simulated replicas of recorded natural conditions.

Earlier studies, as well as future laboratory inquiries into quantitative genetics using exotic or moderately unnatural environments, will still remain of interest. Such studies teach us much about the extent to which natural populations release heritable fitness variation in altered environments, and this is probably a reasonable paradigm for the onset of evolution following environmental change. Furthermore, it may be possible to learn much more about the substructure of fitness (Istock et al. 1976a), about the genetic correlation of fitness characters, and about patterns of genetic determination for fitness characters, using laboratory environments (Mather 1973). We should be able to unravel the processes of selection which increase or decrease $V(a)$ and $\acute{V}(a)$ through long-run laboratory experiments. However, direct testing of hypotheses about polygenic variability for fitness as well as more superficial traits under natural selection will have to be carried out, at least partly, in natural populations. It will be important also to determine the extent to which recent methodological and theoretical initiatives (Harris 1964; Prout 1971a and b; Mather and Jinks 1971; O'Donald 1971; Hill and Nicholas 1974; Thompson 1976; Lande 1976, 1977; Park 1977a and b; Weiss 1979; Gianola 1979a and b) can be adapted to field studies.

The remainder of this section will be a selective review of the quantitative genetics of the primary fitness characters and related considerations. As we shall see, even with the sketchy state of our knowledge, there are suggestions of interesting regularities as well as fascinating patterns.

Since the appearance of Falconer's (1960) text, it has frequently been concluded that fitness characters show less genetic variation than most of the more superficial metric traits. There seems to be a decline of additive genetic variance and heritability as one studies characters closer and closer to the primary fitness characters (Bader 1965; Warburton 1967; Tantawy and El-Helw 1970). One needs to be cautious about this conclusion, however, since most of the supporting data come from dometicated animals which have been, and probably remain, subject to intense directional selection. We know from work with artificial laboratory selection for fast development in *Wyeomyia smithii* that a sizable additive variance, 33 percent of penotypic variance from nature, is at least temporarily eroded to zero in 15 generations or less (Istock et al. 1976a; Istock 1978). We have also observed a significant reduction in the additive variance of development time when selection was imposed in favor of a warm-season diapause and longer development time resulted as a correlated response (Istock et al. 1976a; Istock 1980).

The most important considerations at the moment, however, are the differences in genetic variability to be found: (1) between different primary fitness characters manifested within the same species population, (2) between different species for the same primary fitness trait, and (3) for any of the fitness characters expressed during different segments of the lifespan, particularly the egg, larval, and pupal periods of insects. Bear in mind that we shall usually have to rely on laboratory data with the concomitant uncertainty about how faithfully these data reflect the actual fitness variation in nature.

Variation of Several Fitness Characters at Once

Using wild type, homozygous marker (spineless and aristapedia) and segregating marker strains of *Drosophila melanogaster*, Tantawy and El-Helw (1970) produced an extensive assessment of additive variance for: (1) all three primary fitness characters; (2) durations for subintervals of pre-adult development; (3) one measure of body size (thorax length); and (4) "Productivity," which in their sense includes survival along with the other fitness characters because it measures a one generation, adult to adult, net reproductive rate. The most striking feature of their data, from our present perspective, is the concentration of almost all of the additive fitness variation in egg to adult development time, and principally in the duration of the pupal and larval periods. Longevity and egg production showed little or no heritable variation. Productivity or net reproduction had small overall, though apparently significant, variation. Body size had the highest measure of heritable variation. The high level of additive variance for development time characters could be due to selection which preserves such variation, to the slow release of potential variation, or to the absence of selection for development time. The low level of heritable variation for other fitness traits is presumably the result of past directional selection in the laboratory or in nature, for as we shall see below, fertility and survival frequently are genetically variable in populations of a number of *Drosophila* species, including *D. melanogaster*.

A complex selection experiment with dramatic results was performed by Hollingsworth and Maynard Smith (1955), using isofemale lines of *Drosophila subobscura* derived from wild-caught females. Selection was for either fast or slow development, and all lines were inbred by brother–sister mating. The experiment extends beyond an analysis of development time because correlated changes in fertility and survival ensued. Selection and inbreeding elicited male infertility in all cases, but this result occurred much more rapidly in lines selected for slow development, and in one case the extinction of a slow-selected line was temporarily avoided for a part of the line which was switched to selection for fast development. Hence, the effect of inbreeding was modified by the pattern of selection. The authors concluded that the association of slow development and infertility was due to increased homozygosity, but pleiotropy could also have been involved. Zygotic survival (hatchability) decreased in some lines. In one line heritability for development time increased from 0.22 to 0.89. This rise in heritability was almost certainly a product of linkage disequilibrium engendered by inbreeding, but is quantitatively instructive because it suggests a level of genetic determinism much greater than we will subsequently find. Hollingsworth and Maynard Smith also reported that inbred lines derived from laboratory stocks of the same species show female infertility and no male infertility. The results of Hollingsworth and Maynard Smith defy clear interpretation. One is reminded of Mather and Harrison's (1949) phrase "the manifold effects of selection." It is clear that the genetic analysis of fitness characters and prediction of responses under selection will not be easy.

Variation in Developmental Time

More appears to be known about the genetic variability of developmental rate characters than about the other two primary fitness characters. The question which immediately arises is whether this greater study of development time variations actually reflects a wider occurrence of this type of fitness variation relative to genetic variation for survival and fertility.

Morris and Fulton (1970) repeatedly measured the heritability of the heat requirement for the termination of pupal diapause, morphogenesis of the adult within the pupa, and eclosion from pupa to adult in *Hyphantria cunea* (webworm moth). They found the heritability of pupal heat requirement to be in the range of 0.6–0.8. They measured heat requirement as the number of degree days above 51 degrees F accumulated by an individual pupa prior to eclosion to the adult. This is a correlate of pupal development time. Morris and Fulton interpreted this variation in pupal duration as a genetic adaptation to a temporally variable environment. Environments in different years alter the relation between the timing of the emergence of the moths and the concurrent quality of the apple tree foliage on which they feed. Subsequently, using the expression: ln (offspring heat req.) = 2.456 + 0.602 ln (parent heat req.), and adjustment for successive annual mortality due to temperature variation from year to year, Morris (1971) predicted the sequential change in actual heat requirement after natural selection for those *H. cunea* localities from which 13-year records of observed pupal heat requirement were available. Considering all the sources of error possible in such an attempt, Morris was remarkably successful. He obtained a correlation of 0.75 between observed and predicted values.

Using a foundation population derived from 32 laboratory strains of *Drosophila pseudoobscura*, all but one of which were homozygous for the same chromosomal inversion (arrowhead), Marien (1957) succeeded with artificial selection for fast development. His selection for slow development did not produce a significant change, which is surprising because selection for slow development is usually successful.

Sang and Clayton (1957) observed significant selective advance for both fast and slow larval development with artificial selection applied to a mass mated population of *Drosophila melanogaster* recently colonized from the "wild" (Edinburgh). The selective responses were symmetrical with a realized heritability of 0.21 (realized heritability = character response/selection differential). Sang (1962) later repeated selection for fast and slow development in *D. melanogaster* (composite population of four inbred lines) growing on deficient axenic diets. Here is a clear attempt to force adaptation to atypical environments. On a low-pyroxidine diet, response to selection for fast and slow development was successful and indicated a realized heritability of 0.20. On a low-casein diet there was again a selective response, this time yielding a realized heritability of 0.10. Crosses between the pyroxidine and casein lines indicated that specialization had occurred.

Clarke, Maynard Smith, and Sondhi (1961) also found significant response to selection for both fast and slow development using *Drosophila subobscura*. The responses were asymmetrical with realized heritabilities of 0.063 and 0.186 for the fast and slow lines respectively.

Prout (1962) performed a careful selection experiment with *Drosophila melanogaster*. A foundation stock derived from a wild population in a Southern California citrus grove was divided into lines subject to stabilizing or disruptive mass selection for development time over 40 generations. Control lines were random-mated for development time. The variance in development time decreased with stabilizing selection and increased with disruptive selection. Mean development time declined with stabilizing selection and rose under disruptive selection. Survival from egg to adult fell in the lines subject to disruptive selection, but was unchanged with stabilizing selection. Additive genetic variance decreased with stabilizing selection, but the increase in development time variance with disruptive selection did not involve an increase in additive genetic variance.

Polygenic mutations accumulated on the second chromosome in stocks of *Drosophila pseudoobscura* were used by Mukai and Yamazaki (1971) to estimate the genetic and phenotypic correlation of development time and survival (viability) for both homozygotes and heterozygotes. The mean development time became longer as mutations accumulated in second chromosome homozygotes. Development time became shorter in coupling heterozygotes (mutations on only one of the homologous chromosomes). The correlation between development time and survival was consistently negative, longer development time with reduced survival, with correlation coefficients ranging from −0.4 to −0.9 over both homozygotes and heterozygotes. It was estimated that the average delay in development was 2.5 hours at 25°C for each polygenic mutation. Phenotypic correlations between development time and survival were weaker −0.13 to −0.6. The authors suggested, rather mysteriously, that the primary effect of the mutant polygenes was on viability and that the effect on development time was a pleiotropically derived effect in some sense. The remarkable aspect of this study is the interplay of mutation, fitness character variation, and fitness character correlation which it exposes. Campbell (1962) reported striking variation and correlation of larval development time, egg size, fecundity, and pupal size under the control of variant X chromosomes in the spruce budworm, *Choristoneura fumiferana*.

Studies of the quantitative genetics and adaptive consequences of variation in development time have been done using populations of *Tribolium* species (e.g., Englert and Bell 1970; Scheinberg et al. 1967; Dawson 1965a and b, 1966, 1975, 1977). All of these studies offer convincing evidence from selection experiments that development time is genetically variable in *Tribolium spp*. Dawson has reported realized heritabilities for development time in two different stocks of *T. castaneum* of 0.24 and 0.15. Scheinberg et al. report a value of 0.26. Dawson failed to find selective advance for fast development in a newly colonized "wild" stock. He demonstrated that development time was under stabilizing selection for an intermediate optimum, with genetic variability preserved, in laboratory stocks. He has also argued that density-dependent regulation through cannibalism in either one or two species populations may be the mechanism favoring slower development.

Selection for fast and slow development time was successful with newly colonized stocks (Rochester, New York) of the mosquito *Wyeomyia smithii*, and it was shown that strong correlated responses between development time and warm-season diapause appear no matter which character is actually selected. Fast development was associated with low tendency to diapause and vice versa suggesting pleiotropy. The heritability of development time, with all diapause suppressed, obtained by parent–offspring regression, was 0.33 (Istock et al. 1976a and b; Istock 1978). The variability in diapause is expressed throughout the warm season among the progeny of most wild-caught females when the progeny are tested in the laboratory under the critical photoperiod and diurnal temperature cycles like those of early and late summer at the locality of capture (Istock 1980). The persistence of such variation was explained as an individual adaptation to selection for a fluctuating and, temporally uncertain, intermediate optimum (Istock 1978, 1980). The development rate variability seems to be expressed almost exclusively in the fourth and last larval instar because phenotypic variability is very small in other prereproductive stages, particularly the egg and pupal stages. Diapause occurs in the third instar. The fourth instar is the time when material for egg production is garnered, since the northern mosquitos of this species are incapable of bloodfeeding (Moeur and Istock 1980).

Diapause is an especially interesting facet of development time for studies in evolutionary ecology because it exercises so much influence over the timing and number of generations within and between years, and thereby must be directly involved in adaptive changes in seasonality. In addition to the cases of diapause variability mentioned above, we also know of such genetic variability from studies with northern *Drosophila littoralis* (Oikarinen and Lumme 1979), the pink bollworm, *Heliothis zea* (Herzog and Phillips 1976), the milkweed bug *Oncopeltus fasciatus* (Dingle et al. 1977), and the gypsy moth, *Lymantria dispar* (Lynch and Hoy 1978).

Variability for Survival

Genetic variation for differences in survival probabilities (usually called viabilities in the *Drosophila* literature) is known almost entirely from genetically precise studies with mutant alleles or chromosome variants in *Drosophila* species. From the early work of Dobzhansky (1946) to the present, evidence has accumulated which makes the case for the existence of such fitness variation unassailable (cf. Spiess 1958). Testing of different chromosomes from wild populations of *Drosophila* species almost invariably reveals not only survival variation, but also the superiority or "normality" of heterozygotes (Dobzhansky 1970; Watanabe and Ohnishi 1975). Vetukhiv (1954) and Wallace and Vetukhiv (1955) demonstrated heterozygote superiority in both viability and fecundity for F_1 hybrids from the crossing of different local, natural populations of several *Drosophila* species. These crosses showed F_2 viabilities and fecundities slightly lower than those for within population matings. It would be interesting to know the longer term effects of such experimental introgression of genetic material from geographically separate populations of the same species. The superiority of heterozygotes may hold even when variants produce lethal effects in the homozygous state (Salceda 1967; Anderson 1969).

The surprising thing is that heritable variation for survival differences exists in such abundance. Even if we assume that it is all due to heterosis with consequent genetic segregation and thus give sway to the "balance" school of population genetics (Dobzhansky 1970; Lewontin 1974), we still have to explain the physiological and ecological reasons for heterozygote superiority. One of the most interesting analyses on this point is due to Breese and Mather (1960). In studying the effect on viability of four different segments of chromosome III in *Drosophila melanogaster*, they produced a mapping of the dosage effect upon viability of different parts of the genome. In addition their study allowed comparison of polygenic effects for viability and bristle number along the same chromosome. Breese and Mather found quite uneven effects for viability along the chromosome, and they discovered that there was some tendency for an increasing number of heterozygote chromosome segments to confer higher viability, though heterozygosity per se did not always yield the highest viability. It seems unlikely that any random heterozygote will be superior to every homozygote over all environmental variations encountered in nature.

Artificially induced mutations affecting polygenic traits may not readily show the adaptive advantage of heterozygotes seen with naturally occurring variants. Strickberger (1969) and Ohnishi (1977) did not find a rise in mean fitness in populations of *Drosophila pseudoobscura* and *Drosophila melanogaster* after induction of polygenic mutations. However, the evidence on these points is conflicting. Salceda (1967) found that radiation-induced recessive lethals were maintained at high frequencies in laboratory populations of *Drosophila melanogaster* because of their heterotic effects. Ayala (1967) reported substantial short-term and longer-term (130 weeks) increases in two surrogate fitness measures, population density and biomass as well as a possible increase in longevity in two X-irradiated populations contrasted with a control population of *Drosophila serrata*.

Tobari and Nei (1965) observed that X-irradiation increased the genetic variance for abdominal and sternopleural bristle numbers in *Drosophila melanogaster*. The mean number of bristles remained unchanged, suggesting that the radiation-induced mutations were symmetrical for positive and negative effects on these metric traits. It might be instructive to pursue similar investigations of the effects of artificial mutations on fitness character variation.

Beginnings have been made in the study of the genetics of lifespan. Russell (1957) reported that outbred mice survive months longer than the inbred parental strains from which they are derived. Roberts (1961) found that small strains of mice outlive larger strains by six months. However, Nash and Kidwell (1973) failed to find lifespan differences between inbred and hybrid lines. It might be more promising to study genetic varation for longevity with newly colonized stocks than to study inbred lines, even though one loses the conveniences for genetic analysis which inbred lines offer.

Variability in Fertility

As with survival, it seems paradoxical that genetically controlled differences in offspring production should be preserved under natural selection. Yet selection experiments frequently reveal variation for fertility. Are we merely observing the release of potential variation? I think this may be partly, but not entirely, the explanation. In a remarkable experiment, Doyle and Hunte (1981) held the amphipod crustacean *Gammarus lawrencianus* in culture for 26 generations under conditions allowing continual population growth. After such "free" growth the laboratory stock had survival and fertility rates twice those of a newly colonized stock (6-8 generations of laboratory acclimation) from the same site in nature, and the intrinsic rate of increase was 2.7 times higher. If truly the result of genetic change, and it is difficult to imagine otherwise, these observations raise the question of whether such fitness variation is not mostly the potential fitness variation of the wild population.

A direct assessment of the heritability of clutch size in a great tit (*Parus major*) population in nature (Wytham Woods, England) was achieved by Perrins and Jones (1974). After corrections for variations between years, and between habitats of differing food richness, Perrins and Jones calculated a heritability of about 0.48 for clutch size. They surmised that selection around fluctuating intermediate optima from year to year and habitat to habitat maintains this high level of genetic variability. Previously, Perrins (1964, 1965) found ecological evidence that clutch size might come under selection for intermediate values.

Further information about the amount of heritable variation for fertility comes mostly from studies with domesticated animals, particularly chickens, mice, and pigs (cf. Nordskog et al. 1967), and from studies with *Drosophila* species.

Gains in the selective improvement of the egg production of domestic chickens may have ceased in recent years, though improvement in body weight continues (Nordskog 1977). Here we may have an example of the uncoupling of a superficial character from a fitness character under artificial selection.

Most estimates of the heritability of litter size in mice range from lows near zero to highs of about 0.25 (Falconer 1960; Rahnefeld et al. 1962; Boylan et al. 1961). An unusually high value of 0.40 was reported by Dadlani and Prabhu (1970). They also found negative genetic correlations for the age of mother and litter size and the weight of mother and litter size (-0.62 and -0.47, respectively). Estimates of the heritability of litter size in swine range from 0 to 0.24 with 0.15 a typical value (Craft 1958; Falconer 1960). Craig and Chapman (1953) and Jinks and Broadhurst (1963) have reported high heritability values in the range of 0.40 to 0.50 for litter size in rats.

As a correlated response during selection for high and low chaeta number in *Drosophila melanogaster* populations, Mather and Harrison (1949) observed drastic declines in fertility. Though probably the result of inbreeding, this correlated response demonstrates the previous existence of genetic variation for fertility. After a period of relaxed selection Mather and Harrison were able to continue selection for chaeta number without further effects on fertility. Robertson's (1957) estimate for the heritability of egg production by *Drosophila melanogaster* was 0.20. Kearsey and Kojima (1967) found that all three major chromosomes in *Drosophila melanogaster* contribute to variability in egg hatchability, but the second and third chromosomes command more variability than the X chromosome. Dominance deviations toward high hatchability predominated, while in contrast body weight was largely controlled by additive effects. Tait and Prabhu (1970) obtained an estimate of 0.68 for the heritability of egg production using *Drosophila melanogaster* and regression of daughters on mothers. They also found that the heritability was dramatically reduced to values as low as 0.07 when the second chromosome was held constant and the background of other chromosomes varied. Tait and Prabhu also obtained a value of 0.82 for the heritability of egg hatchability in one cross, and again varying the background caused reductions of this value. Richardson and Kojima (1965) could readily select for egg production in outcrossed *Drosophila pseudoobscura*. Gowen (1952) and Robertson and Reeve (1955) demonstrated that mean egg production increased with the number of heterozygous chromosomes. Additive and nonadditive genetic variance for egg production exerted by the X chromosome of *D. melanogaster* was studied by Chapco (1977). He found that segments of the X chromosome had additive influences in homozygous condition, but displayed dominance in the heterozygote. Different X chromosome segments showed significantly different effects on egg production. It should be possible to map the dosages for a fitness character associated with a given part of a chromosome. If this were done for two or more different fitness characters we would know something about the extent of intertwining, or interspersion, of the polygenic determinants of fitness characters inside the genetic material.

Potential variation for litter size was vividly exposed in Falconer's (1971) selection for larger litter size after an apparent plateau in selective advance had been reached. After reaching the plateau at about generation 23, Falconer continued selection for higher litter size with inbreeding in nine lines. Four of the inbred lines survived and were crossed 11 generations later. A substantial gain beyond the earlier apparent selection limit occurred by generations 36 to 45.

Heritable variation among genes influencing fertility may serve to create partial reproductive isolation between geographically isolated populations. This type of polygenic variation was presumably the source of the high degree of male hybrid sterility found in crosses between *Drosophila pseudoobscura* from Bogota,

Columbia and from localities in Central and North America (Prakash 1972; Dobzhansky 1974). The extent of genetic variability in male fertility is largely unstudied.

Discussion

Potential polygenic variation for the primary fitness characters is widely present in the animal populations studied to date. However, this kind of heritable variation is not universally present, and when it is, its amount and heritability are not uniform from character to character or species to species. The taxonomic array for which we have estimates of polygenic variation in fitness traits is small. Gains in understanding should come by exending this array. When present, potential fitness variation is moderately heritable, with heritabilities in the range of 0.05 to 0.40. This amount of fitness variation is higher than frequently assumed, and it has been repeatedly shown to support substantial response to artificial selection. Two objectives for future theoretical and empirical studies will be the explanation of the definite tendency to intermediacy in the heritability of fitness characters, and the adaptive consequences, including both ecological specialization and speciation, of these modest but selectable levels of genetic variance.

Actual polygenic variation for the primary fitness characters, the amount expressed in nature, is poorly known. Three studies offer direct indications of the amounts and kinds of actual variation (Morris 1971; Perrins and Jones 1974; Istock 1978). It is likely that heritable fitness differences within natural populations are common, though this variation may be of a kind which does not immediately create directional selection for higher intrinsic fitness in accord with Fisher's theorem of natural selection. Instead, it appears that actual fitness variation may be present only for some fitness traits, particularly when it supports adaptation under selection for a fluctuating intermediate optimum in developmental timing, or in fertility under a fluctuating selection regime and density-dependence. Under such selection regimes fitness

Heritable Variation

variation may be conserved and it might even accumulate. Clearly we need more field studies of the expression and fate of polygenic variation.

The relation between potential and actual polygenic variation is unclear. In the first place, we have no good studies of the environmental lability of quantitative traits, though we typically assume that genetic variances and heritabilities vary wildly under environmental change. The important issue is to measure the response of the additive variance in fitness characters throughout a range of variation actually experienced by the natural population. Clearly, direct measurement in the field, when achievable, is best, and standard methods of offspring-parent regression and sib correlation may be feasible under some field circumstances.

Potential variation of both the "hidden" and "environmentally releasable" kinds is probably abundant, and perhaps periodically builds up through some kind of accumulation mechanism. In this way the processes of micro and macroevolution could be joined, though as with all genetic variation each population will carry a specialized profile of variation and its derivative probabilities of extinction. It surprises no one to say that current selection models cannot accurately portray macroevolution.

Despite the abstractions of quantitative genetics it is clear from many studies that we can study the genetic architecture (Mather 1973) of fitness characters in relation to the chromosomes, or genome. The yawning hiatus between quantitative and molecular genetics will be at least partially bridged when both parts of genetics can make much clearer statements about the relation between chromosomes and their respective domains of genetic analysis. Variation in the pattern of polygenic influence along different "wild-caught" versions of the same chromosome will add another and possibly important dimension to the study of genetic architecture and the genetic variation of natural populations.

Partial and tentative answers can be given to some of the questions posed earlier, here restated in shorter form.

1. How do we think the genes behave in a statistical sense?
2. How are superficial and fitness traits coupled genetically and developmentally?

3. How heritable is the variation for fitness traits?
4. How much is actual polygenic variation? How much is potential polygenic variation?
5. How do stabilizing selection, mutation and genetic recombination mold fitness variation?
6. What happens when stabilizing selection and directional selection on different fitness characters conflict?
7. How do the amounts and limits of genetic variation for fitness traits influence the evolution of life-history patterns?

How Do the Genes Behave?

The additivity assumption of quantitative genetics theory is an adequate model for many studies of the polygenic inheritance of fitness characters, but dominance effects may predominate in some cases. Because of the multifarious relations of physiology with development, survival, and reproduction we might anticipate that fitness characters will offer the "largest" polygenic systems available for study, and there may be a kind of statistical safety in size here. However, to unravel the role of specific gene actions underlying metric traits, fitness characters will probably be less useful than more restricted morphological traits.

How Are Superficial Traits Connected?

We seem to have little solid information to address this question. Surely the color patterns of cryptics and mimics, pheromones, courtship patterns and mate selection, body size, habitat selection, migration and dispersal, aggregation, sexual dimorphism, and color or other morphological polymorphisms are at one time or other, in one species or other, tied to fitness. Such traits will be of overriding importance for survival and reproduction during periods of major alteration by directional selection and macroevolution. Studies such as that of Breese and Mather (1960) reveal the interspersion of polygenic systems for different traits, but we are still too ignorant of the genetic architecture involved and the extent of interspersion to say how the polygenes of primary fitness characters and the genetic determinants of more superficial traits coevolve. Heterotic influences have repeatedly been noted for their effects on metric traits, but no general conclusion about them is possible on present information.

How Heritable Are Fitness Traits?

Developmental characters tend to have heritabilities of 0.10 to 0.40 with possibly much higher values on occasion. Fertility measures are typically lower in heritability with values of 0.05 to 0.25. Variation for survival seems to be commonplace, if one is allowed to generalize from the *Drosophila* studies. Such variation remains somewhat mysterious because we would expect it to be pressed to very low levels by universal directional selection. Breese and Mather (1960) did find that the dominance effects on viability were all unidirectional toward higher survival, while with chaeta number the dominance effects went toward both higher and lower values of the trait.

How Much Is Expressed in Nature?

The meager data available suggest that selective forms of fitness variation may be routinely expressed in a natural population. As yet we do not have a single measurement of the amount of additive or total genetic variance for a fitness character in nature.

Stabilizing Selection and Fitness Variation

Following Lande's theoretical work, Morris's study with *Hyphantria cunea*, and our studies of *Wyeomyia smithii*, there is some reason to believe that stabilizing natural selection for an intermediate optimum requires, and to a degree conserves, polygenic variation for fitness. Under Lande's conception mutation would serve to restore losses from selection and presumably genetic drift. We need an increased empirical grasp of the effects of polygenic mutation on fitness.

Directional Selection and Stabilizing Selection in Conflict

Laboratory selection experiments in great number make it clear that the polygenic variation released by a natural population will often support the directional microevolution of fitness traits. Eventually selective advance reaches a plateau and the additive variance may be measurably zero (Istock et al. 1976a). Hence it is clear that Fisher's theorem can have its predicted result.

Correlated responses in selection experiments and significant genetic correlations among different fitness traits are found often enough to suggest that interspersion of fitness polygenes is sometimes significant. But we don't know if the observed correlations are the result of selection molding the underlying genetic architecture to be this way or if such correlations are merely unavoidable genetic and physiological constraints. To the extent that different fitness characters can reside in separate parts of the genetic system, stabilizing selection for one characteristic could co-occur with directional selection for another.

When genetic systems are interlocked either stabilizing selection or directional selection must predominate. For example, strong stabilizing selection for development time might both preserve developmental variation and engender variation in survival, which would otherwise be reduced to near zero by directional selection. This view is opposite to the one suggested by Mukai and Tamazaki (1971).

Life History Consequences of Fitness Variation

Fitness characters are largely synonymous with life history features in the standard demographic representation of life history phenomena. Directional selection will shape life history features toward a fixed end. In the absence of recurrent waves of new variation for the fitness characters a population would fail to evolve new adaptations to prevailing or altered environments. At least some life history features are probably variable in most populations.

Variation observed for development time in a number of species is particularly interesting in a life history context, for example the patterns of heritable variation for larval and pupal development reviewed above. The stage of development where genetic variability is found is itself variable among species. In *Hyphantria* the variation was pupal, in *Drosophila melanogaster* it was largely for the pupal stage but also substantial for larval development time, and in *Wyeomyia* the genetic variation was for larval development time and not at all in the pupal stage. The developmental variation may even be largely in the fourth instar for northern *Wyeomyia*. Another mosquito, *Aedes triseriatus*, seems to have variation spanning weeks or months for the duration of the egg stage even when covered with water, though we don't know how much of this is genetic (Istock, unpublished observations).

Hairston et al. (1970) contended that natural selection could not simultaneously increase both survival and fertility. Their argument was purely ecological under the assumption of constant density-dependent regulation of populations and a necessary tradeoff between energy requirements for survival and reproduction. Genetic studies of fitness characters argue that there might be genetic constraints which also limit multiple improvements in fitness. However, caution is needed here because we have found experimentally that it is possible to select for faster development in *Wyeomyia smithii* without loss in either survival or fertility (Istock et al. 1976a). Admittedly, our experimental populations were not constantly under density-dependent limitation, nevertheless free genetic recombination and occasional lapses of density-dependence might allow the evolution of the sort of "superorganism" which Hairston et al. thought impossible.

All of this amounts to a modest beginning when one reflects on the largeness of Stearns' (1976) call for an understanding of the genetic limits to the life history variation. Increased effort in the study of fitness and life history variation will be worth the candle if it contributes to an evolutionary synthesis of Mendelian, quantitative and ecological genetics.

Summary

Since the early development of systematic approaches to the study of evolution, two types of genetic analysis have coexisted within population genetics. These are the Mendelian and biometrical modes of analysis. The differences between the two views of inheritance have been resolved and it is clear that both ways of approaching genetic and evolutionary problems are fruitful.

The biometrical approach is particularly important for the study of the primary fitness characters, survival, development time, and fertility because these characters typically show continuous variation due to underlying polygenic determination.

Even when directional selection in the past has caused the loss of all additive genetic variance for fitness in Fisher's sense, there may remain an additive variance based on deviations in fitness characters arising from changing effects of genes with varying genetic background from individual to individual. With selection for an intermediate optimum, and mutation, polygenic variation may persist indefinitely. An adaptive role for such enduring variation may involve response to a temporally or spatially fluctuating and uncertain environment in nature.

A distinction must be made between the "actual" polygenic variation expressed in nature and the total "potential" polygenic variation manifested upon complete uncovering of variation through recombination and through growth of the population in many different environments. Much of the variability in fitness characters which we see in laboratory studies may reflect potential rather than actual amounts and ranges of variation. Heritabilities of .05 to .40 are typical for laboratory determinations of variability in fitness characters.

In the future, we need to estimate the actual polygenic variation directly in natural populations. Only a tiny number of such preliminary attempts currently exist.

It should also be possible to study the interspersion of the polygenes for several primary fitness characters together along the chromosomes (genome) after the fashion of earlier studies. Conflicts may develop when one fitness character is subject to stabilizing selection and another to directional selection.

Development time appears to be the most variable of the primary fitness traits. It may frequently be subject to selection for an intermediate optimum in both natural and laboratory populations. The developmental stage at which heritable variation persists seems to vary considerably from species to species. Such patterns may reflect what past evolution has done to shape the distribution of variability for life-history phenomena. At present, we cannot interpret these patterns because we do not know how the amounts and limits to variation in fitness characters are involved in life-history evolution.

ACKNOWLEDGMENTS

I am particularly grateful to Ernst Caspari, Uzi Nur, Julia Graham, and William J. Etges for their criticism of this paper. Wyatt Anderson, Peter Dawson, Timothy Prout, Thomas Caraco, Rodney Norman, John Jaenike, Loy Merkle, Elizabeth Maisel, John Moeur, Kathryn Marshall, and Joseph Warnick all helped me by discussing various parts of this paper, and I thank them. My research work included herein, and the opportunity to write this paper were made possible by NSF grant DEB-7724615.

REFERENCES

Anderson, W. W. 1969. Genetics of natural populations. XLI. The selection coefficients of heterozygotes for lethal chromosomes in *Drosophila* on different genetic backgrounds. *Genetics* 62:827–836.

Ayala, F. 1967. Evolution of fitness. III. Improvement of fitness in irradiated populations of *Drosophila serrata*. *Genetics* 58:1919–1923.

Bader, R. S. 1965. Heritability of dental characters in the house mouse. *Evolution* 19:379-384.

Boag, P. R. and P. R. Grant. 1978. Heritability of external morphology in Darwin's finches. *Nature* 274:793-794.

Boylan, W. J., W. E. Rempel, and R. E. Comstock. 1961. Heritability of litter size in swine. *J. Anim. Sci.* 3:566.

Breese, E. L. and K. Mather. 1960. The organization of polygenic activity within a chromosome in *Drosophila*. II. Viability. *Heredity* 14:375-399.

Bulmer, M. G. 1971. The effect of selection on genetic variability. *Am. Nat.* 105:201-211.

Campbell, I. M. 1962. Reproductive capacity in the genus *Choristoneura* Led. (Lepidoptera: Torticidae). I. Quantitative inheritance and genes as controllers of rates. *Can. J. Genet. and Cytol.* 4:272-288.

Chapco, W. 1977. Correlations between chromosome segments and fitness in *Drosophila melanogaster*. I. The X-chromosome and egg production. *Genetics* 85:721-732.

Clarke, J. M., J. Maynard Smith, and K. C. Sondhi. 1961. Asymmetrical response to selection for rate of development in *Drosophila subobscura*. *Genet. Res. Camb.* 2:70-81.

Craft, W. A. 1958. Fifty years of progress in swine breeding. *J. Anim. Sci.* 17:960.

Craig, J. V. and A. B. Chapman. 1953. Experimental test of predictions of inbred line performance in crosses. *J. Anim. Sci.* 12:124-139.

Crow, J. F. and M. Kimura. 1970. *Introduction to Population Genetics Theory*. New York: Harper and Row.

Dadlani, H. V. and S. S. Prabhu. 1970. Quantitative genetic studies in the mouse (*Mus musculus* L.). I. Litter size. *J. Genetics* 60:184-191.

Darlington, C. D. and K. Mather. 1949. *The Elements of Genetics*. London: Allen and Irwin.

Dawson, P. S. 1965a. Genetic homeostasis and developmental rate in *Tribolium*. *Genetics* 51:873-885.

Dawson, P. S. 1965b. Estimates of components of phenotypic variance for developmental rate in *Tribolium*. *Heredity* 20:403-417.

Dawson, P. S. 1966. Developmental rate and competitive ability of *Tribolium*. *Evolution* 20:104-116.

Dawson, P. S. 1975. Directional versus stabilizing selection for developmental time in natural and laboratory populations of flour beetles. *Genetics* 80:773-783.

Dawson, P. S. 1977. Evolutionary adjustment of developmental time in mixed populations of flour beetles. *Nature* 270:340-341.

Dingle, H., C. K. Brown, and J. P. Hegmann. 1977. The nature of genetic variance influencing photoperiodic diapause in a migrant insect. *Oncopeltus fasciatus. Am. Nat.* 111:1047-1059.

Dobzhansky, Th. 1946. Genetics of natural populations. XIII. Recombination and variability in populations of *Drosophila pseudoobscura. Genetics* 31:269-290.

Dobzhansky, Th. 1970. Genetics of the Evolutionary Process. New York: Columbia University Press.

Dobzhansky, Th. 1974. Genetic analysis of hybrid sterility within the species *Drosophila pseudoobscura. Hereditas* 77:81-88.

Doyle, R. W. and W. Hunte. 1981. Demography of an estuarine amphipod (*Gammerus lawrencianus*) experimentally selected for high "r": A model of the genetic effects of environmental change. *Can. J. Fish. Aquatic Sci.* 38:1120-1127.

East, E. M. 1910. A Mendelian interpretation of variation that is apparently continuous. *Am. Nat.* 44:65-82.

Englert, D. C. and A. E. Bell. 1970. Selection for time of pupation in *Tribolium castaneum. Genetics* 64:541-552.

Falconer, D. S. 1960. *Introduction to Quantitative Genetics*. New York: Ronald Press.

Falconer, D. S. 1971. Improvement of litter size in a strain of mice at a selection limit. *Genet. Res. Camb.* 17:215-235.

Feldman, M. W. and R. C. Lewontin. 1976. The heritability hangup. *Science* 190:1163-1168.

Fisher, R. A. 1918. The correlations between relatives on the supposition of Mendelian inheritance. *Trans. Roy. Soc. Edinburgh* 52:399-433.

Fisher, R. A. 1958. *The Genetical Theory of Natural Selection*. New York: Dover.

Gelderman, H. 1975. Investigation on inheritance of quantitative characters in animals by gene markers. *Theor. Appl. Genet.* 46:319-330.

Gianola, D. 1979a. Estimation of genetic covariance from joint offspring-parent and sib-sib data. *Genetics* 93:1039-1049.

Gianola, D. 1979b. Heritability of polychotomous characters. *Genetics* 93:1051-1055.

Gowen, J. W. 1952. Hybrid vigor in *Drosophila*. In J. W. Gowen, ed. *Heterosis*. Ames: Iowa State College Press.

Hairston, N. G., D. W. Tinkle, and H. M. Wilbur. 1970. Natural selection and the parameters of population growth. *J. Wildlife Mgmt.* 34:681-689.

Harris, D. 1964. Expected and predicted progress from index selection involving estimates of population parameters. *Biometrics* 20: 46-72.

Herzog, G. A. and J. R. Phillips. 1976. Selection for a diapause strain of the bollworm, *Heliothis zea. J. Heredity* 67:173-175.

Hill, W. G. and F. W. Nicholas. 1974. Estimation of heritability by both regression of offspring on parent and intraclass correlation of sibs in one experiment. *Biometrics* 30:447-468.

Hollingsworth, M. J. and J. Maynard Smith. 1955. The effects of inbreeding on rate of development and on fertility in *Drosophila subobscura. J. Genetics.* 53:295-314.

Istock, C. A. 1978. Fitness variation in a natural population. In H. Dingle ed. *Evolution of Insect Migration and Diapause.* New York: Springer-Verlag.

Istock, C. A. 1980. Natural selection and life history variation: theory plus lessons from a mosquito. In R. F. Denno and H. Dingle eds. *Species and Life History Patterns.* New York: Springer-Verlag.

Istock, C. A., S. Wasserman, and H. Zimmer. 1975. Ecology and evolution of the pitcher-plant mosquito. 1. Population dynamics and laboratory responses to food and population density. *Evolution* 29:296-312.

Istock, C. A., J. Zisfein and K. J. Vavra. 1976a. Ecology and evolution of the pitcher-plant mosquito. 2. The substructure of fitness. *Evolution* 30:535-547.

Istock, C. A., K. J. Vavra, and H. Zimmer. 1976b. Ecology and evolution of the pitcher-plant mosquito. 3. Resource tracking in a natural population. *Evolution* 30:548-557.

Jinks, J. L. and P. L. Broadhurst. 1963. Diallel analysis of litter size and body weight in rats. *Heredity* 18:319-336.

Kearsey, M. J. and B. W. Barnes. 1970. Variation for metrical characters in *Drosophila* populations. *Heredity* 25:11-21.

Kearsey, M. J. and Ken-Ichi Kojima. 1967. The genetic architecture of body weight and egg hatchability in *Drosophila melanogaster. Genetics* 56:23-37.

Lande, R. 1976. The maintenance of genetic variability by mutation in a polygenic character with linked loci. *Genet. Res. Camb.* 26: 221-235.

Lande, R. 1977. The influence of the mating system in the maintenance of genetic variability in polygenic characters. *Genetics* 86:485-498.

Lande, R. 1980. The genetic covariance between characters maintained by pleiotropic mutations. *Genetics* 94:203-215.

Lewontin, R. C. 1974. *The Genetic Basis of Evolutionary Change.* New York: Columbia University Press.
Lewontin, R. C. 1977. The relevance of molecular biology to plant and animal breeding. In E. Pollak, O. Kempthorne, and T. B. Bailey eds. *Int'l Conf. Quant. Genetics.* Ames, Iowa State Univ. Press.
Linney, R. B., W. Barnes and M. J. Kearsey. 1971. Variation for metrical characters in *Drosophila* populations. *Heredity* 27:163-174.
Lynch, C. B. and M. A. Hoy. 1978. Diapause in the gypsy moth: environment specific mode of inheritance. *Genet. Res. Camb.* 32:129-133.
Manning, A. 1963. Selection for mating speed in *Drosophila melanogaster* based on the behavior of one sex. *Anim. Behav.* 11:116-123.
Marien, D. 1956. Selection for developmental rate in *Drosophila pseudoobscura. Genetics* 50:3-15.
Mather, K. 1973. *Genetical Structure of Populations.* London: Chapman and Hall.
Mather, K. and B. J. Harrison. 1949. The manifold effect of selection. *Heredity* 3:1-52 and 131-162.
Mather, K. and J. L. Jinks. 1971. *Biometrical Genetics.* London: Chapman and Hall.
Maynard Smith, J. 1979. The effects of normalizing and disruptive selection on genes for recombination. *Genet. Res. Camb.* 33:121-128.
Moeur, J. E. and C. A. Istock. 1980. Ecology and evolution of the pitcher-plant mosquito. 4. Larval influence over adult reproductive performance and longevity. *J. Anim. Ecol.* 49:775-792.
Morris, R. F. and W. C. Fulton. 1970. Heritability of diapause intensity in *Hyphantria cunea* and related fitness responses. *Can. Ent.* 102:927-938.
Morris, R. F. 1971. Observed and simulated changes in genetic quality in natural populations of *Hyphantria cunea. Can. Ent.* 1103:893-906.
Mukai, T. and T. Yamazaki. 1971. The genetic structure of natural populations of *Drosophila melanogaster.* X. Developmental time and viability. *Genetics* 69:385-398.
Nash, D. J. and J. F. Kidwell. 1973. A genetic analysis of lifespan, fecundity, and weight in the mouse. *J. Heredity* 64:87-90.
Nilsson-Ehle, H. 1909. *Keruzungsuntersuchungen an Hafer und Weizen.* Lunds Universitets Arsskrift n. s., ser 2, vol. 5, no. 2.
Nordskog, A. W. 1977. Success and failure of quantitative genetic theory in poultry. In E. Pollak, O. Kempthorne and T. B. Bailey eds. *Proc. Intl. Cong. Quant. Genetics.* Ames, Iowa State Univ. Press.
Nordskog, A., W. M. Festing, and M. W. Verghese. 1967. Selection for

egg production and correlated responses in the fowl. *Genetics* 55:179-191.

O'Donald, P. 1971. Natural selection for quantitative characters. *Heredity* 27:137-153.

Ohnishi, O. 1977. Spontaneous and methanesulfonate-induced mutations controlling viability in *Drosophila melanogaster*. III. Heretozygous effect of polygenic mutation. *Genetics* 87:545-556.

Oikarinen, A. and J. Lumme. 1979. Selection against photoperiodic reproductive diapause in *Drosophila littoralis*. *Hereditas* 90:119-125.

Park, Y. C. 1977a. Theory for the number of genes affecting quantitative characters. I. Estimation of and variance of the estimation of gene number for quantitative traits controlled by additive genes having equal effect. *Theor. Appl. Genet.* 50:153-161.

Park, Y. C. 1977b. Theory for the number of genes affecting quantitative characters. II. Biases from drift, dominance, inequalty of gene effects, linkage disequilibrium and epistasis. *Theor. Appl. Genet.* 50:163-172.

Parsons, P. A. 1974. Male mating speed as a component of fitness in *Drosophila*. *Behav. Genetics* 4:395-404.

Perrins, C. M. 1964. Survival of young swifts in relation to brood size. *Nature* 201:1147-1148.

Perrins, C. M. 1965. Population fluctuations and clutch size in the great tit, *Parus major*. *J. Anim. Ecol.* 34:601-647.

Perrins, C. M. and P. J. Jones. 1974. The inheritance of clutch size in the great tit (*Parus major* L.). *Condor* 76:225-229.

Pollak, E., O. Kempthorne, and T. B. Bailey eds. 1977. *Proceedings of the International Conference of Quantitative Genetics*. Ames: Iowa State University Press.

Prakash, S. 1967. Association between mating speed and fertility in *Drosophila robusta*. *Genetics* 57:655-663.

Prakash, S. 1972. Origin of reproductive isolation in the absence of apparent genetic differentiation in a geographic isolate of *Drosophila pseudoobscura*. *Genetics* 72:143-155.

Prout, T. 1962. The effects of stabilizing selection on time of development in *Drosophila melanogaster*. *Genet. Res. Camb.* 3:364-382.

Prout, T. 1971a. The relation between fitness components and population prediction in *Drosophila* I: The estimation of fitness components. *Genetics* 68:127-149.

Prout, T. 1971b. The relation between fitness components and population prediction in *Drosophila* II: Population prediction. *Genetics* 68:151-167.

Provine, W. B. 1971. *The Origins of Theoretical Population Genetics.* Chicago, University of Chicago Press.

Rahnefeld, G. W., W. J. Boylan and R. E. Comstock. 1962. Genetic correlations between growth rate and litter size in mice. *Can. J. Genet. Cytol.* 4:272-288.

Richardson, R. H. and Ken-Ichi Kojima. 1965. The kinds of genetic variability in relation to selection responses in *Drosophila* fecundity. *Genetics* 52:583-598.

Roberts, R. C. 1961. The lifetime growth and reproduction of selected strains of mice. *Heredity* 16:369-381.

Robertson, F. W. 1954. Studies in quantitative inheritance. V. Chromosome analysis of crosses between selected and unselected lines of different body size in *Drosophila melanogaster. J. Genetics* 52:494-520.

Robertson, A. 1955. Selection in animals: synthesis. *Cold Spring Harbor Symp. Quant. Biol.* 20:225-229.

Robertson, F. W. 1957. Studies in quantitative inheritance. XI. Genetic and environmental correlation between body size and egg production in *Drosophila melanogaster. J. Genet.* 55:428-433.

Robertson, F. W. and E. C. R. Reeve. 1955. Studies in quantitative inheritance. VIII. Further analysis of heterosis in crosses between inbred lines of *Drosophila melanogaster. Z. Indukt. Abstamm. u. VererbLehre* 86:439-458.

Russell, E. S. 1957. A study of lifespan and pathology tendencies of breeding mice from ten inbred strains. *Proc. Am. Assn. Can. Res.* 2:245.

Salceda, V. 1967. Recessive lethals in second chromosomes of *Drosophila melanogaster* with radiation histories. *Genetics* 57:691-699.

Sang. J. H. 1962. Selection for rate of development using *Drosophila melanogaster* cultured axenically on deficient diets. *Genet. Res. Camb.* 3:90-103.

Sang, J. H. and G. A. Clayton. 1957. Selection for larval development time in *Drosophila. J. Heredity* 48:265-270.

Scheinberg, E., A. E. Bell, and V. L. Anderson. 1967. Genetic gain in populations of *Tribolium castaneum. Genetics* 55:69-90.

Sinnott, E. W., L. C. Dunn, and Th. Dobzhansky. 1958. *Principles of Genetics.* New York, McGraw-Hill.

Slatkin, M. 1970. Selection and polygenic characters. *Proc. Natl. Acad., Sci. USA* 66:87-93.

Spiess, E. B. 1958. Effects of recombination on viability in *Drosophila. Cold Spring Harbor Symp. Quant. Biol.* 23:239-250.

Stearns, S. C. 1976. Life-history tactics: A review of the ideas. *Quart. Rev. Biol.* 51:3-47.

Strickberger, M. W. 1969. Factors determining rates of the evolution of fitness in laboratory populations of *Drosophila pseudoobscura*. *Genetics* 62:639-651.

Tait, W. M. and S. S. Prabhu. 1970. Fecundity and hatchability in *Drosophila melanogaster* with a second chromosome held unchanged. *J. Genetics* 60:152-158.

Tantawy, A. O. and M. R. El-Helw. 1970. Studies on natural populations of *Drosophila*. IX. Some fitness components and their heritabilities. *Genetics* 64:79-91.

Thoday, J. M. 1977. Effects of specific genes. In E. Pollak, O. Kempthorne, and T. B. Bailey eds. *Int'l Conf. Quant. Genetics*, Ames, Iowa State Univ. Press.

Thompson, R. 1976. Design of experiments to estimate heritability when observations are available on parents and offspring. *Biometrics* 32: 283-304.

Tobari, I. and M. Nei. 1965. Genetic effects of X-rays on quantitative characters in a heterogeneous population of *Drosophila melanogaster*. *Genetics* 52:1007-1015.

Vetukhiv, M. 1954. Integration of the genotype in local populations of three species of *Drosophia*. *Evolution* 8:241-251.

Wallace, B. and M. Vetukhiv. 1955. Adaptive organization of the gene pools of *Drosophila* populations. *Cold Spring Harbor Symp. Quant. Biol.* 20:303-309.

Warburton, F. E. 1967. Increase in the variance of fitness due to selection. *Evolution* 21:197-198.

Watanabe, T. K. and S. Ohnishi. 1975. Genes affecting productivity in natural populations of *Drosophila melanogaster*. *Genetics* 80:807-819.

Weiss, V. 1979. The heritability of difference scores when environments are correlated. *Biom. J.* 21:171-177.

Wright, S. 1968. *Evolution and the Genetics of Populations*. Volume 1. *Genetic and Biometric Foundations*. Chicago, University of Chicago Press.

4 Measuring Genetic Variation in Natural Populations: Where Are We?

JEFFREY R. POWELL

Quantifying levels of genetic variation in natural populations remains one of the outstanding problems in experimental population genetics, and indeed in all of biology. A definitive solution to this problem will affect, in a very basic way, our view of how evolution occurs and consequently our understanding of the whole organic world. The relevance of the question to population biology is obvious, one need only note that essentially every article in this symposium in one way or another addresses this problem, though most do so indirectly. I propose to address the problem directly. I shall review the most recent developments in this field, and then provide some speculations as to the future. Finally, I hope to very briefly point out that such pivotal questions in population biology may have far-reaching implications for some very basic practical human problems. As will become apparent, we are a long way from solving the problem, although new techniques are getting us closer to a solution.

Population Geneticists Discover Molecular Biology

In Richard Lewontin's (1974) by now classic book, *The Genetic Basis of Evolutionary Change*, the history of the problem of measuring levels of genetic variation was thoroughly and elegantly reviewed. Here, I shall review the progress made since that book appeared. Doubtless this will be an idiosyncratic and biased review as it will emphasize the work with which I am most familiar, that of my collaborators and myself.

The major preoccupation of experimental population geneticists for the last 15 years has been to apply techniques and principles of molecular biology to the study of genetic variation in populations. The technique of gel electrophoresis has proven to be the most practical technique and the species studied now number in the hundreds (reviewed by Powell 1975a Nevo 1978). Generally, several soluble enzymes whose activity can be visualized on a gel, are surveyed for electrophoretically detectable variation among many individuals. The genetic basis of the variation is confirmed by crosses. Or, more commonly, assumptions are made about the number of loci being surveyed and the allelic nature of the variants. While different species reveal various levels of variation in such studies, generally invertebrates prove to be polymorphic at about 50 percent of the loci studied and an individual is heterozygous at 10-20 percent of its loci on average; vertebrates tend to be somewhat less variable. If one wishes to use these estimates as even approximate measures of the variation in the genome as a whole, then certain criteria must be met (Lewontin and Hubby 1966). The two most important (and testable) are (1) all or nearly all allelic substitutions at a locus must be detectable; and (2) the loci studied must represent a random sample of the genome. Probably the most important advance in the last five years is the realization that neither of these criteria is met in the vast majority of studies done to date. Ironically, by not meeting criterion (1) we may have underestimated the level of variation and by not meeting criterion (2) we may have overestimated. This is probably true of both intra- and interspecific comparisons.

"Hidden" Variation Revealed

As was clear from the very beginning (Lewontin and Hubby 1966) electrophoresis as it is usually employed in allozyme surveys reveals only a fraction, an *unknown* fraction, of amino acid substitutions. Finding just how much variation was hidden has been something of a surprise. The patterns of variation may also vary depending upon how much variation is revealed.

Singh et al. (1976) list a series of techniques which may be employed to detect amino acid substitutions in polypeptides; no single study has to date employed all techniques. I would like to discuss two studies which employed techniques different from those usually used with rather spectacular results.

The works of Singh et al. (1976) and Coyne (1976) on *Drosophila pseudoobscura* and *D. persimilis* illustrate how misleading usual electrophoretic studies can be. They employed a technique of sequential electrophoresis. By studying lines made homozygous for the locus coding for xanthine dehydrogenase (Xdh) they could make multiple electrophoretic runs under varying conditions. They varied the pH of the gel buffer, acrylamide concentration, and also looked for heat stability differences. To summarize, they found five to six times the number of alleles than were revealed by single condition electrophoresis. Furthermore, the genetic differences between the Bogotá subspecies and N. American *D. pseudoobscura* greatly increased, as did the differences between the sibling species. Thus the remarkable increase in intraspecific variation was paralleled by any equally impressive intertaxa increase in genetic differentiation. This technique of multiple pass electrophoresis is thought to detect almost all ($\approx 85\%$) amino acid substitutions (Ramshaw et al. 1979).

A caveat is necessary at this point. Finnerty and Johnson (1979) and Finnerty et al. (1979) have accumulated evidence that electrophoretic mobility differences in *Drosophila* xanthine dehydrogenase may be due to post-transcriptional or post-translational modification caused by loci other than *Xdh*. The whole subject remains rather controversial and I don't wish to dwell on this further (see arguments in Letters to the Editor, *Genetics* [1979] 92 [1 and 2]). This controversy, however, does highlight some important problems with the multiple-condition electrophoresis technique. In order to be confident that the variation being detected is due to allelic variants at the locus coding for the enzyme it is necessary to map the variants to the location of the locus. In order to do such genetic studies it is necessary to be able to determine the phenotype of individual organisms. It may be possible to use the genetic advantages of *Drosophila* to do the required studies although this has not yet been done for any locus. However with most organisms it would be very laborious and costly, if not impossible, to analyze genetically all the variants revealed. Of course this problem has not inhibited allozyme surveys of refractory organisms; the genetic interpretation of gel phenotypes is usually based on homology to organisms for which genetic data are available. Unfortunately the non-allelic mobility effects were detected in the organism for which genetic studies are the easiest.

The possible importance of post-transcriptional or post-translational modification is illustrated in figure 4.1. Considerable diversity of a protein can be generated by several loci with a few alleles each rather than a single locus with a great number of alleles. The possible mechanisms maintaining the diversity would be different in the two circumstances. For example, maintaining more than about ten alleles at a single locus by heterosis is very difficult (Lewontin et al. 1978).

The second study I wish to discuss was done in our laboratory and employed the technique of isoelectric focusing (IEF), a technique which has empirically been shown to be generally more powerful in separating proteins than is electrophoresis (Hayes and Wellner 1969; Spencer and King 1971; Milkman and Koehler 1977; Rico 1979; Powell, unpublished data; see however Ramshaw and Eanes 1978, for contrary results). In our original electrophoretic survey of the *Drosophila willistoni* group we compared the chromosomal and allozymic variation of South American and Caribbean Island populations of *D. willistoni* (Ayala, Powell, and Dobzhansky 1971). The chromosomal inversions indicated founder effects on the islands while the allozymic variation on islands was very similar both qualitatively and quantitatively to that found on the continent. The question Rico (1979) addressed was whether the similarities between islands and the continent were artifacts arising from the failure of starch gel electrophoresis to detect differences which actually exist. Two Brazilian and one Puerto Rican populations of *D. willistoni* were surveyed for variation at eleven loci, those which proved to give good resolution

Genetic Variation in Natural Populations 101

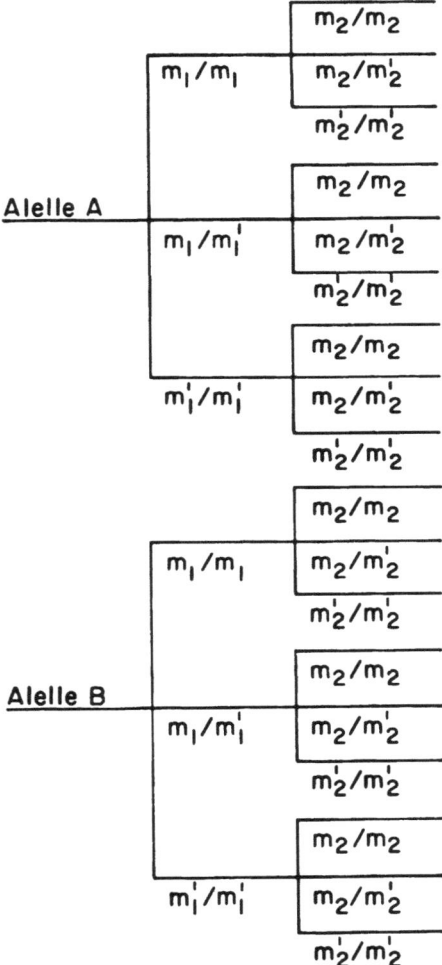

Figure 4.1. Scheme whereby three loci with two alleles each can generate what appears to be a single locus with 18 alleles.

Note: Alleles A and B code for amino acid sequences of a protein. Two loci, m_1, and m_2, are assumed to modify the protein; homozygotes and heterozygotes modify the protein differently.

in IEF gels. A summary of Rico's study is presented in table 4.1. Clearly one's view of these populations is influenced by the techniques used.

Table 4.1. Average Heterozygosity in *D. willistoni* Caribbean and S. American Populations as Revealed by Isoelectric Focusing and Starch Gel Electrophoresis.

	Isoelectric Focusing[a]	Starch Gel Electrophoresis[b]
Mayagüez, Puerto Rico	11.9 ± 3.3	16.2 ± 0.6
Brazil: São Carlos	23.4 ± 6.8	18. ± 0.8
Rio Grande del Sul	27.3 ± 10.5	

[a]From Rico (1979)
[b]From Ayala, Powell, and Dobzhansky (1971). Based on six Caribbean Islands and four continental populations surveyed for 28 loci.

Are Enzymes a Biased Sample of the Genome?

That enzyme-coding loci are a biased sample of the genome, at least in *Drosophila*, was hinted at in a much ignored paper by Berger and Weber (1974). They studied ribosomal proteins in strains of *D. melanogaster* from around the world. Of the 60 to 70 protein spots they could identify, only one varied in one strain. They used both one- and two-dimensional polyacrylamide electrophoresis. They speculated that genes coding for ribosomal proteins may be a particularly conservative class of loci because of "steric restrictions imposed by membership in a three-dimensional organellar array" and the importance of ribosome function in the survival of the organism.

Brown and Langley (1979) were the first to attack the problem in *Drosophila* using the two-dimensional technique of O'Farrell (1975). They extracted proteins using sonication and an extraction buffer which should solubilize even membrane-bound proteins. Isoelectric focusing was done in the first dimension and SDS acrylamide electrophoresis in the second. As an "internal standard" they used flies which were heterozygous over standard balanced inversions for the second and third chromosomes. Of the estimated 54 loci (really protein spots) only 6 varied among the twenty independent genomes they studied. Racine and Langley (1980) have extended such studies to mice with very similar results.

McConkey et al. (1979), and Walton et al. (1979) have applied similar methods to the study of human tissue cultures. They used double-labeling techniques where one of the two lines run together was labeled with tritium and the other with ^{14}C. Both studies, one with 4 and the other with 5 cell lines, reached similar conclusions. Of the several hundred spots which were resolved only about 2 percent varied, and the average heterozygosity was estimated at less than 1 percent. The lines studied represented at least two "races," Blacks and Caucasians; thus one might predict this would maximize the amount of variation revealed.

Similar studies on the *Drosophila obscura* group have yielded comparable preliminary results. We have not yet used the two-dimensional technique but we have done both IEF and SDS polyacrylamide electrophoresis in one dimension followed by general protein strains. In the IEF studies about ten adults per isofemale strain were in each lane; 44 readable bands were detected by Coomassie blue staining. In the electrophoresis studies we used a much more sensitive protein stain (Switzer et al. 1979) which allows the visualization 72 or 73 proteins from a single fly. In the IEF studies, 44 reliable bands were visualized. Table 4.2 summarizes our results to date. These data, although preliminary, are reasonably consistent with the studies just cited. Polymorphism at the loci coding for these proteins is about one fifth that for

Table 4.2. Preliminary Results of Studies on Variation in General Proteins from Several Inbred Lines of *Drosophila*.

Comparison	Number of Strains Compared	Bands Different
IEF; Coomassie Blue		
Within N. American *pseudoobscura*:	10	5/44
Between N. Amer. and Bogotá *pseudoobscura*	10:3	5/44
Between N. Amer. *pseudoobscura* and *persimilis*	10:1	9/44
SDS, Silver Stain:		
Within N. Amer. *pseudoobscura*:	3	3/73
Between N. Amer. and Bogotá *pseudoobscura*	3:1	2/73
Between *pseudoobscura* and *persimilis*	3:1	4/72
Between *pseudoobscura* and *subobscura*	3:1	9/73

[a] The *D. pseudoobscura* lines came from a wide variety of geographic sources.

enzyme-coding loci. Most interesting are the intertaxa results. The partially reproductively isolated subspecies of *D. pseudoobscura bogatana* is no more different from North American populations than are North American populations from one another. Again this contrasts with enzyme-coding loci, which have been shown to have diverged quite substantially between these subspecies (Ayala and Dobzhansky 1974; Singh et al. 1976; Powell 1979a). The relative conservation of these proteins compared with enzymes is highlighted by the *pseudoobscura/suboscura* comparison. These species have different most common alleles at about 80 percent of their enzyme-coding loci (Lakovaara et al. 1972). Since the strains studied here were reasonably inbred one would expect them to have become fixed for different alleles at the majority of their loci, assuming enzyme-coding loci are a random sample of all loci. Yet these general protein stains show only 12 percent differences. Thus it appears that relative to enzyme-coding loci, the loci coding for these proteins are more conservative both within and between taxa.

Why should there be this difference? In addition to the reasoning of Berger and Weber (1974) mentioned above, Brown and Langley (1979) pointed out another possible bias in the allozyme studies. They showed that many of the most commonly studied enzymes were known to be polymorphic before they were applied to population studies. The simple reason for this was that published techniques for staining *Drosophila* enzymes concentrated on those for which one could do genetic analysis—i.e., were variable.

To summarize this section, the developments in the last 5 years have once again thrown open the question of levels of genetic variation in natural populations. Are enzymes a biased sample of the genome? Or is it possible that the proteins detected by general stains are a biased sample? Presumably the proteins detected in such studies are those in highest concentration in the organism. In terms of levels of genetic variation is there something peculiar about proteins in high concentration? The studies of McConkey et al. (1979) revealed up to 600 protein spots. Are even the 600 most common proteins a biased sample? Also, if the preliminary results hold, what will these new studies mean for our view of the genetics of speciation? Is there less of a "genetic revolution" than indicated by allozyme studies? These are the kinds of questions it will be necessary to consider in the future if we are to really measure levels of variation within and between taxa for this class of gene loci. Results during the last 5 years have made it increasingly clear just how far away we are from having definitive answers.

Variation in the Other 99 Percent of the Genome

Elsewhere, I have calculated that less than 1 percent of a eukaryotic genome codes for the amino acid sequence of proteins (Powell 1975b). If we really want to measure levels of genetic variation in populations, it seems obvious we must also be concerned with this other 99 percent of the genome. In order to do this it will be necessary to find phenotypes whose expression is controlled by this other part of the genome; it must also be possible to enumerate these phenotypes in samples of individuals from natural populations. What possible function could this part of the DNA serve so that we might know what phenotypic differences might indicate genetic variation in this part of the genome? At least some of this genetic material must be involved with gene regulation, a subject of considerable recent speculation among evolutionists (Wilson 1976; Wallace 1975). While the speculation has been stimulating, what do we really know about gene regulation as it pertains to population biology and evolution? Some data have been accumulated recently which bear on this question.

In our laboratory we have been concentrating on the α-amylase gene-enzyme system in *Drosophila*. In addition to there being structural gene polymorphism (allozymes) there is a reasonably well-characterized gene regulation polymorphism controlling the tissue-specific expression of amylase (Abraham and Doane 1978; Powell and Lichtenfels 1979). This latter polymorphism is for patterns of amylase along the length of the adult midgut. Differences among individuals in their amylase activity pattern are at least to a large degree under genetic control, presumably some kind of regulatory elements or "genes." The main question we have asked is how do these two polymorphisms behave with respect to one another. Table 4.3 summarizes some of our results

Table 4.3. Percentages of Amylase Midgut Activity Patterns and Amylase Allozymes in 5 Populations.

Midgut Activity Pattern		D. pseudoobscura			D. persimilis	
AMG	PMG	Flagstaff	Zirahuen	Bogotá	Fish Creek	Mather
123	12	3	—	3	92	92
123	10	42	49	35	—	7
123	00	3	2	—	—	—
120	10	28	21	8	—	—
103	10	10	10	30	—	—
103	00	3	2	8	—	—
100	10	6	9	5	8	1
100	00	3	2	8	—	—
Other		2	5	3	—	—
Amylase Allozymes						
S		17	74	100	5	6
F		83	26	—	82	40
VF		—	—	—	14	53

SOURCES Powell (1979) and Powell et al. (1980).

(see Powell 1979b; and Powell et al. 1980a, for more extensive data). The most striking and consistent aspect of these results is the evolutionary independence of the two types of polymorphism. For example *D. pseudoobscura* populations from Flagstaff, Arizona and Zirahuen, Mexico have very similar frequencies of midgut patterns but differ greatly in frequencies of amylase allozymes. The Bogotá population is fixed for an *Amy* allele but is just as variable for midgut patterns. *D. persimilis* exhibits a converse situation; it is nearly fixed for a particular midgut activity pattern but is about as polymorphic as *D. pseudoobscura* for the *Amy* allozymes.

The relevance of these results to the ascertainment of levels of genetic variation is twofold. First, it is clear there is considerable variation in the part of the genome controlling amylase expression; we have detected 13 different patterns in natural populations of *D. pseudoobscura*. The question which must be asked is how widespread is this type of variation? It is impossible to give a definite answer at this time, partially because of our ignorance of how genes are regulated in eukaryotes. Different workers have different concepts of what constitutes variation in gene regulation.

Nevertheless, I believe there already exists enough information to predict that naturally occurring variants in the expression of enzyme-coding loci are widespread, perhaps even more so than allozyme variation. While it is true that negative data are often left unpublished, with few exceptions whenever anyone has made a reasonably thorough search for variation in level of enzyme expression and/or variation in tissue specificity, it is found. The only systematic search for gene regulation is that of Laurie-Ahlberg et al. (1980). Six of seven presumably randomly chosen enzymes in *Drosophila melanogaster* were polymorphic in their levels of expression; five of the seven had activity variants unlinked to the structural gene. Since our ability to detect such variation is rather insensitive, I believe these results can be taken as evidence that such variation is widespread. Of course there remains the question of whether looking at variation in enzyme expression rather than proteins in general is a biased measure of the level of gene regulation variation.

The second point of relevancy of the data in table 4.3 is the evident independence of allozyme and gene regulation variation. This implies that whatever the evolutionary/ecological forces are that control the frequencies of the enzyme variants, they are not the same factors that control the frequencies of polymorphic genetic factors responsible for the tissue-specific expression of the enzyme. In other words, from an evolutionary/ecological perspective as well as a purely genetic standpoint, the type of variation revealed by these studies on gene regulation is different in kind from that revealed by allozyme studies. Such variation may be even more relevant to population biology than the allozymes have been.

Another possible function of the 99 percent of the genome not coding for amino acid sequences of proteins is control of recombination rates. In *Drosophila melanogaster* the amount of heterochromatin in the X chromosome affects the rate of recombination of both the X chromosome and autosomes (Yamamoto and Miklos 1978; Yamamoto 1979). Control of recombination, especially intragenic recombination is one of the few viable explanations of the evolutionary role of intervening sequences (e.g., Gilbert 1980). Certainly the presence of intervening sequences makes more plausible the schema of Watt (1972), who proposed that the generation of genic diversity by intragenic recombination could occur at rates much higher than "standard" mutation rates (figure 4.2). This could also explain the apparent contradiction of the correlation of the size of an enzyme and its level of variation; in humans there appears to be no correlation (Harris et al. 1977) while in *Drosophila* there is a positive correlation (Koehn and Eanes 1977). If intragenic recombination is a significant force generating genic diversity, then the lengths of intervening sequences may be as, or more, important as is the length of the amino acid sequence.

The evolutionary significance of rates of generation of genic diversity and control of recombination is clearly great. These phenomena play a central role in several unsolved problems in population biology—e.g., the selectionist/neutralist controversy of protein polymorphism and the evolution of sex. I might also point out that the polymorphisms which have probably played the largest role in shaping our thinking about natural population genetics are ones whose essential effect is to control recombination, namely paracentric inversions in *Drosophila*. Besides chromosome structure variation which controls recombination, there is evidence of naturally occurring genetic variation for recombination rates unrelated to gross chromosome variation (e.g., Kidwell 1972; Chinnici 1971). Future research into the evolutionary role of re-

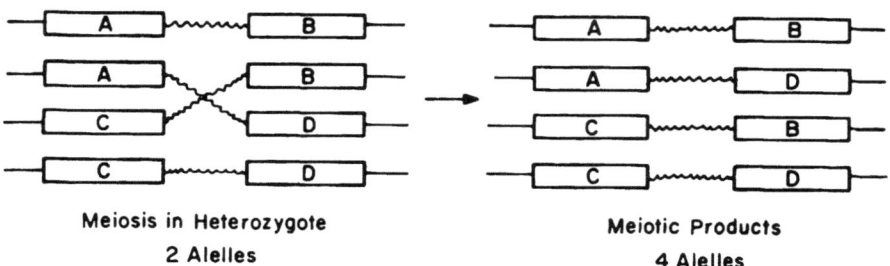

Figure 4.2. **Example of intragenic recombination generating new allelic variation.**
Note: The wavy area represents the intervening sequence and the boxed areas are the protein-coding regions. The different letters represent different coding sequences for that part of the protein.

combination, especially as it is related to the new discoveries about the structure of genomes, will be very exciting and fruitful. Unlike research on protein polymorphisms, there can be no doubt that recombination rates do have an adaptive significance.

While it may be unsatisfying, it is also possible that much of the 99 percent of the genome not involved in coding sequences has no function whatsoever. The discovery of "dead genes," nontranscribed DNA sequences with clear homology to functioning genes, implies that eukaryotic genomes often carry evolutionary relics. Many geneticists, myself included, have resisted the notion that evolution could be so inefficient as to carry along such excess baggage. We may have been suffering under the same misconception as proponents of optimization theory in ecology. Reality may not conform to the best of all possible alternatives.

Is Population Biology of Any Use?

As with many areas of so-called basic research, it was difficult to predict at the outset what practical value would accrue from the discipline we call population biology. I believe the field has now developed to the point where the principles and techniques developed by population biologists can now be applied to some important problems facing our species. The cost to benefit ratio is, or soon will be, less than one.

As an example of a practical use of detecting genetic diversity, I will discuss a continuing project dealing with *Aedes aegypti*, the yellow fever mosquito (Tabachnick and Powell 1978, 1979; Tabachnick et al. 1979; Powell et al. 1980b). We have studied more than 40 populations for electrophoretically detectable variation at 22 enzyme-coding loci. These populations came from throughout the pantropical and subtropical distribution of the species. One question we have asked is whether the genetic differentiation among regions is great enough to allow diagnosis of the geographic origin of a population based on allozyme frequencies. Figure 4.3 summarizes the results of a multivariate discriminant analysis based on allozyme frequencies. These first two canonical

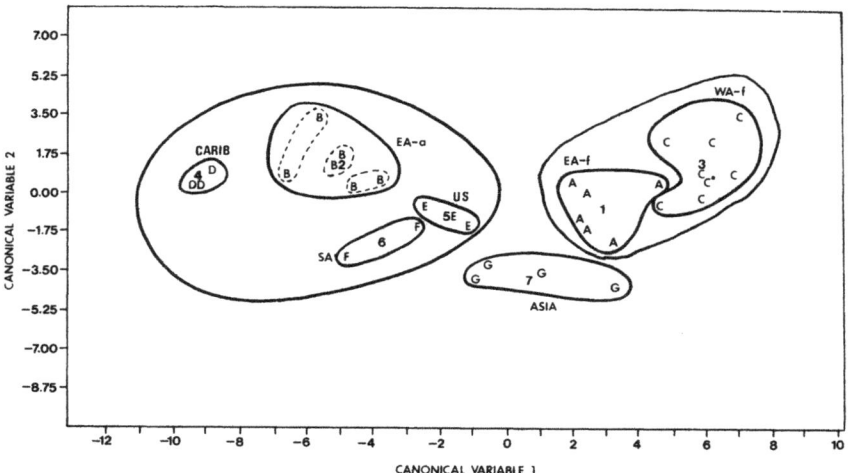

Figure 4.3. Multivariate discriminate analysis of allozyme frequencies in *Aedes aegypti*, the yellow fever mosquito.

Note: A's are population samples of the *formosus* subspecies from East Africa. B's are *aegypti* subspecies from East Africa. C's are the *formosus* subspecies from West Africa. D's are Caribbean samples. E's are Southwest U.S. samples. F's are S. American samples. G's are Asian samples. The two larger inclusions represent the subspecies as described by Mattingly (1957). The dashed lines include two samples from the same locations taken one or two years apart. The C* means two samples were graphically indistinguishable. The numbers are the center for that subgroup. (From Powell et al. 1980.)

variables represent 92 percent of the total variation in allele frequencies. The ability to diagnose the geographic origin of *A. aegypti* is especially important in light of the evidence that there is genetic variation for the efficiency at which strains from various regions can transmit viruses (Aitken et al. 1977; Beaty and Aitken, 1979). The fact that Asia has apparently never experienced yellow fever has led some authors to speculate that the genotypes which colonized this continent have a reduced vectorial capacity (Dudley 1934).

The recognition that such vector-borne diseases represent a three-tiered co-evolutionary system will be valuable in understanding and controlling these number one killers of humans. The genetic diversity at each level is just now becoming appreciated. Similar diversity in agricultural and forest pests is also being recognized (e.g., Anderson et al. 1979).

Conclusions

Progress is being made in the measurement of genetic variation in natural populations. Much remains to be learned. I feel confident that with the new molecular biology of the eukaryotic genome, when all the data are in our view of the meaning of genetic variation and its causes will be quite different from what it is today. Despite all this progress and confidence in the future, a central nagging problem remains. How does all this variation at the molecular level relate to the kind of variation which leads to adaptive traits? Is any of what I've discussed relevant to the kinds of variation so often discussed in this symposium, traits like life history parameters and sexual versus parthenogenetic reproduction? Certainly the genetic variation in such obviously adaptive traits has a molecular genetic basis, but the question is how to measure it. It may be argued that since we have not solved the problem of the adaptive significance of the rather limited amount of variation revealed by allozyme surveys, how will we ever be able to solve the problem for all the additional kinds of variation? I would argue the converse. It may be that the relative crudeness of the usual electrophoretic techniques has been the cause of the problem. What we need are new data, more accurate data, data on all kinds of genetic variation. When we have a better understanding of the molecular structure and function of eukaryotic genomes and the variation inherent therein, we shall be in a much better position to relate the variation to Darwinian evolution.

As researching academic population biologists, it is the nature of our profession to be very critical of our work and to constantly point to the limitations of our knowledge. However it is also occasionally worthwhile to contemplate just how much we have really accomplished. We have established certain principles and techniques which have stood the test of time. Some of these can be of enormous value to human society and have only just begun to be applied to such problems as control of insect pests, evolution of parasites, and conservation of endangered species. If we can make society more aware of the potential value of population biology, then I believe we have a very bright future.

REFERENCES

Abraham, I. and W. W. Doane. 1978. Genetic regulation of tissue-specific expression of amylase structural genes in *Drosophila melanogaster. Proc. Natl. Acad. Sci.*, USA 75:4446-4450.

Aitken, T. H. G., W. G. Downs, and R. Shope. 1977. *Aedes aegypti* strain fitness for yellow fever virus transmission. *Am. J. Trop. Med. Hyg.* 26:985-989.

Anderson, W. W., C. W. Berisford, and R. H. Kimmich. 1979. Genetic differences among five populations of the southern pine beetle. *Ann. Ent. Soc. Am.* 72:323-327.

Ayala, F. J., J. R. Powell, and Th. Dobzhansky. 1971. Polymorphisms in continental and island populations of *Drosophila willistoni. Proc. Natl. Acad. Sci.* USA 68:2480-2483.

Beaty, B. J. and T. H. G. Aitken. 1979. *In vitro* transmission of yellow fever virus by geographic strains of *Aedes aegypti. Mosq. News* 39:232-238.

Berger, E. M. and L. Weber. 1974. The ribosomes of Drosophila. II. Studies on interspecific variation. *Genetics* 78:1173-1183.

Brown, A. J. L. and C. Langley. 1979. Reevaluation of genic heterozygosity in natural populations of *Drosophila melanogaster* by two-dimensional electrophoresis. *Proc. Natl. Acad. Sci.*, USA 76:2381-2384.

Chinnici, J. P. 1971. Modification of recombination frequency in Drosophila. I. Selection for increased and decreased crossing-over. *Genetics* 69:71-83.

Coyne, J. 1976. Lack of genic similarity between two sibling species of *Drosophila* as revealed by varied techniques. *Genetics* 84:593-607.

Dudley, S. F. 1934. Can yellow fever spread into Asia? *J. Trop. Med. Hyg.* 37:273-278.

Finnerty, V. and G. Johnson. 1979. Post-transcriptional modification as a potential explanation of high levels of enzyme polymorphism: xanthine dehydrogenase and aldehyde oxidase in *Drosophila melanogaster. Genetics* 91:695-722.

Finnerty, V., M. McCarron, and G. Johnson. 1979. Gene expression in *Drosophila*: post-translational modification of aldehyde oxidase and xanthine dehydrogenase. *Molec. Gen. Genet.* 172:37-42.

Gilbert, W. 1980. Introns and exons: playgrounds of evolution. In Press.

Harris, H., D. A. Hopkinson, and Y. H. Edwards. 1977. Polymorphism and the subunit structure of enzymes: a contribution to the neutralist-selectionist controversy. *Proc. Natl. Acad. Sci.* USA 74:698-701.

Hayes, M. B. and D. Wellner. 1969. Microheterogeneity of L-amino acid oxidase. *J. Biol. Chem.* 244:6634-6644.
Kidwell, M. G. 1972. Genetic change of recombination value in *Drosophila melanogaster*. I. Artificial selection for high and low recombination-modifying genes. *Genetics* 70:419-432.
Koehn, R. K. and W. F. Eanes. 1977. Subunit size and genetic variation of enzymes in natural populations of *Drosophila. Theoret. Pop. Biol.* 11:330FF.
Lakovaara, S., A. Saura, and C. T. Falk. 1972. Genetic distance and evolutionary relationships in the *Drosophila obscura* group. *Evolution* 26:177-184.
Laurie-Ahlberg, C. C., G. Maroni, G. C. Bewley, J. C. Lucchesi, and B. S. Weir. 1980. Quantitative genetic variation of enzyme activities in natural populations of *Drosophila melanogaster. Proc. Natl. Acad. Sci.* USA 77:1073-1077.
Lewontin, R. C. 1974. *The Genetic Basis of Evolutionary Change*. New York: Columbia University Press.
Lewontin, R. C., L. Ginzburg, and S. Tuljapurkar. 1978. Heterosis as an explanation for large amounts of genic polymorphism. *Genetics* 88:149-170.
Lewontin, R. C. and J. L. Hubby. 1966. A molecular approach to the study of genic heterozygosity in natural populations. II. Amount of variation and degree of heterozygosity in natural populations of *Drosophila pseudoobscura. Genetics* 54:595-609.
McConkey, E. H., B. J. Taylor, and D. Phan. 1979. Human heterozygosity: A new estimate. *Proc. Natl. Acad. Sci.*, USA. 76:6500-6504.
Milkman, R. and R. Koehler. 1977. Isoelectric focusing of MDH and 6-PGDH from *E. coli* of diverse natural origins. *Biochem. Genet.* 14:517-522.
Nevo, E., 1978. Genetic variation in natural populations: patterns and theory. *Theoret. Pop. Biol.* 13:121-177.
O'Farrell, P. H. 1975. High resolution two-dimensional electrophoresis of proteins. *J. Biol. Chem.* 250:4006-4021.
Powell, J. R. 1975a. Protein variation in natural populations of animals. *Evol. Biol.* 8:79-119.
Powell, J. R. 1975b. Isozymes and non-Darwinian evolution: A reevaluation. *Isozymes* 4:9-26, C. L. Markert, ed., Academic Press.
Powell, J. R. 1979a. More genic divergence of the Bogotá population of *Drosophila pseudoobscura. Genetics* 91:s99.
Powell, J. R. 1979b. Population genetics of Drosophila amylase. II.

Geographic patterns and correlations in *D. pseudoobscura*. *Genetics* 92:613-622.

Powell, J. R. and J. M. Lichtenfels. 1979. Population genetics of Drosophila amaylase. I. Genetic control of tissue-specific expression in *D. pseudoobscura*. *Genetics* 92:603-612.

Powell, J. R., M. Rico, and M. Andjelkovic. 1980a. Population genetics of Drosophila amylase. III. Interspecific variation. *Evolution* 34: 209-13.

Powell, J. R., W. J. Tabachnick, and J. Arnold, 1980b. Genetics and the origin of a vector population: *Aedes aegypti*: A case study. *Science*, 208:1385-1387.

Racine, R. R. and C. H. Langley. 1980. Genetic heterozygosity in a natural population of *Mus musculus* assessed by using two-dimensional electrophoresis. *Nature* 283:855-857.

Ramshaw, J. A. M., and W. F. Eanes. 1978. Study of the charge state models for electrophoretic variation using isoelectric focusing of esterase-5 from *Drosophila pseudoobscura*. *Nature* 275:68-70.

Ramshaw, J. A. M., J. A. Coyne, and R. C. Lewontin. 1979. The sensitivity of gel electrophoresis as a detector of genetic variation. *Genetics* 93: 1019-1037.

Rico, M. 1979. Electrophoretically silent genetic variation in the *Drosophila willistoni* group. Ph.D. Dissertation, Yale University.

Singh, R., R. C. Lewontin and A. Felton. 1976. Genetic heterogeneity within electrophoretic "alleles" of xanthine dehydrogenase in *Drosophila pseudoobscura*. *Genetics* 84:609-629.

Spencer, E. M. and T. P. King. 1971. Isoelectric heterogeneity of bovine plasma albumin. *J. Biol. Chem.* 246:201-208.

Switzer, R. C., C. R. Merril, and S. Shifrin. 1979. A highly sensitive stain for detecting proteins and peptides in polyacylamide gels. *Anal. Biochem.* 98:231-237.

Tabachnick, W. J. and J. R. Powell. 1978. Genetic structure of East African populations of *Aedes aegypti*. *Nature* 272:535-537.

Tabachnick, W. J. and J. R. Powell, 1979. A world-wide survey of genetic variation in the yellow fever mosquito, *Aedes aegypti*. *Genet. Res.* 34:215-229.

Tabachnick, W. J., L. E. Munstermann, and J. R. Powell. 1979. Genetic distinctness of sympatric forms of *Aedes aegypti*. *Evolution* 33:287-295.

Wallace, B. 1975. Gene control mechanisms and their possible bearing on the neutralist-selectionist controversy. *Evolution* 29:193-202.

Walton, K. E., D. Styer, and E. I. Gruenstein. 1979. Genetic polymorphism in normal human fibroblasts as analyzed by two-dimensional polyacrylamide gel electrophoresis. *J. Biol. Chem.* 259:7951–7960.

Watt, W. B. 1972. Intragenic recombination as a source of population genetic variability. *Am. Nat.* 106:737–753.

Wilson, A. C. 1976. Gene regulation in evolution. In F. J. Ayala, ed. *Molecular Evolution.* Sunderland, Massachusetts: Sinauer Press.

Yamamoto, M. 1979. Interchromosomal effects of heterochromatin deletions on recombination in *Drosophila melanogaster. Genetics* 93:437–448.

Yamamoto, M. and G. L. G. Miklos, 1978. Genetic studies on heterochromatin in *Drosophila melanogaster* and their implications for the functions of satellite DNA. *Chromosoma* 66:71–98.

5 The Biochemical and Physiological Bases of Aminopeptidase-I (Lap) Polymorphism in *Mytilus edulis*

RICHARD K. KOEHN

Attempting to understand the maintenance of enzyme polymorphism within natural populations is a classic problem in ecological genetics. Resolution of this problem requires us to identify, simplify, and understand the complex interactions among genotypes, phenotypes, and ecological variations. Such interactions, as they specifically pertain to enzyme polymorphism, require an investigative approach different from studies of the environmental genotype interactions that are translated through visible phenotypes such as banding patterns or color polymorphisms. After all, visible markers are visible. Indeed, it has been the genetic basis of color and pattern variation that has had to be established in studies of the evolutionary significance of these polymorphisms. In contrast, genetic polymorphism of enzymes is ubiquitous in populations, but we are virtually ignorant of the phenotypic effects of different enzyme genotypes.

The allele frequencies of polymorphic genes are often statistically correlated with spatial or temporal variation of environmental factors (cf, Hedrick et al. 1976; Endler 1977), but the causal relationship between the population frequency of an enzyme allele and the environmental factors(s) cannot usually be demonstrated. Hence, in order to understand the basis of enzyme polymorphism in nature, we are simultaneously challenged to elucidate the existence of phenotypic variation (beyond the electrophoretic phenotype), "map" detected phenotypes onto measured environmental variables, and to infer from this mapping (or the results of devised tests) both the existence and nature of a functional relationship between genotype and the environment (cf., Clarke 1975; Koehn 1978).

While the existence of biochemical variation among genotypes is not in and of itself evidence for selectively important enzyme polymorphism (Koehn 1978), biochemical differences among enzyme morphs have been described. For example, differences in K_m were reported for phosphoglucomutase variants in *Drosophila melanogaster* (Fucci et al. 1979) and temperature differences in the K_m's of lactate dehydrogenase morphs in a fish have been described by Merritt (1972). However, it is not clear what the adaptive significance of these differences might be. Place and Powers (1979) have correctly pointed out that K_m is not a rate constant and it is upon reaction rate that natural selection would be expected to act. Differences in "activity" or V_{max} have been described for esterase morphs in a fish (Koehn 1969) and glucose-6-phosphate dehydrogenase and 6-phosphogluconate dehydrogenase in *D. melanogaster* (Bijlsma and van der Meulen-Bruijns 1979), but the physiological consequence of variable reaction rate of these enzymes is not known.

The alcohol dehydrogenase polymorphism in *D. melanogaster* has received much attention. The two common alleles differ in activity (Rasmuson et al. 1966; Gibson 1970), a difference that has been attributed to catalytic efficiency (K_{cat}) (Day et al. 1974). There is evidence that the activity differences adapt the flies to levels of environmental alcohol (Vigue and Johnson 1973; Bijlsma-Meeles and van Delden 1974; McDonald and Avise 1976), though the substrate affinities of the enzyme are complex (David et al. 1976). The control of alcohol dehydrogenase activity is also complex, involving gene loci other than the structural locus (Ward and Hebert, 1972; McDonald et al. 1977; McDonald and Ayala 1978).

Place and Powers (1979) have demonstrated temperature dependent differences in K_{cat}/K_m ratios among lactate dehydrogenase genotypes in the Common Killifish, *Fundulus heteroclitus*. The authors convincingly argue that these differences are of evolutionary significance, because a large K_{cat}/K_m is advantageous. Each morph is most frequent in latitudes at which it is the more efficient catalyst under prevailing regimes of temperature.

The foregoing studies emphasize that in addition to allele frequency changes among populations and biochemical properties of enzyme variants, two additional kinds of information are needed: the physiological consequences of biochemical variation and how these consequences may change in relation to particular ecological circumstances.

It is within this context that studies of the differentiation among populations of the marine bivalve, *Mytilus edulis*, have been undertaken. In the following pages, I hope to demonstrate how the combination of biochemical, physiological, and ecological information can elucidate evolutionary mechanisms that are involved in producing genetic differences within and between populations of this species. It is a detailed study of a single locus in a single species, and it also serves to illustrate one investigative approach to the understanding of enzyme polymorphism.

Aminopeptidase-I Variation in *Mytilus Edulis*

Mytilus edulis, like many marine invertebrates, has an extended larval dispersal stage (Bayne 1976). During this developmental stage, larvae are passively dispersed by water currents, which should produce thorough mixing of the offspring of spatially proximate populations. Large genetic differences among populations are not expected.

Of the many gene loci that have been studied in *M. edulis* (Ahmad et al. 1977), allele frequencies at most polymorphic loci exhibit only small differences among populations (Levinton and

Koehn 1976). However, the frequency of alleles at the *Lap* locus differ markedly between populations. The polymorphism of the enzyme product of this gene, termed aminopeptidase-I (Young et al. 1979), has been extensively investigated in phylogenetically diverse marine bivalves (Koehn and Mitton 1972; Ayala et al. 1973; Schaal and Anderson 1974; Buroker et al. 1975; Levinton 1975; Levinton and Fundiller 1975; Murdock et al. 1975; Koehn et al. 1976; Milkman and Koehn 1977; Levinton and Suchanek 1978; Lassen and Turano 1978; Levinton and Lassen 1978a, 1978b; Koehn et al. 1980a, 1980b). The products of the *Lap* gene are detected as "naphthylamidase" activity. Functional similarities have not been established for the naphthylamidase polymorphisms among most studied species, though the "Lap-I" and the "Lap-II" loci reported by many authors probably correspond to the aminopeptidase-I and peptidase-II enzymes respectively, in *Mytilus edulis* (Young et al. 1979).

Spatial variations in the frequencies of *Lap* alleles are typically correlated with variations of salinity and/or temperature (Koehn et al. 1976). In both Europe (Theisen 1978) and North America (Koehn et al. 1976), steep allele frequency clines connect genetically differentiated populations. In North America, clines are generally associated with the entrances to estuaries in waters south of Cape Cod and in more oceanic waters in the environs of Cape Cod (Koehn et al. 1976).

Several studies have focused on a steep cline connecting Atlantic Ocean populations, where the frequency of the Lap^{94} allele is about 0.55, with populations in Long Island Sound, where Lap^{94} equals 0.15 (Koehn et al. 1976; Milkman and Koehn 1977; Lassen and Turano 1978; Levinton and Lassen 1978a; 1978b). At the eastern entrance of Long Island Sound there is a smooth, but steep diminution in the frequency of Lap^{94} with increasing distance into Long Island Sound (figure 2, in Koehn 1979). Lassen and Turano (1978) suggested temporal and spatial instability of this cline and attributed certain deficiencies of heterozygotes in the youngest mussels to population mixing, or the Wahlund effect (Li 1969). The role of natural selection was dismissed in favor of hydrographically maintained population isolation.

The correlation of *Lap* allele frequencies with environmental variables is not unlike that observed in other studies of geographic variation (Hedrick et al. 1976). Although such observations have often been used to argue the case for the action of natural selection in population differentiation, such data do not provide evidence that the gene under study is the target of inferred selection. After all, there has never been disagreement among evolutionary biologists on the role of natural selection in population variation and differentiation; the disagreement (Lewontin 1974) centers upon *whether or not phenotypic differences among enzyme morphs are under the control of natural selection.*

Correlations between the frequencies of enzyme alleles and environmental variables are important in hypothesis formation. For example, the correlations in *M. edulis* between *Lap* allele frequencies and environmental salinity suggest the possibility that the product of this gene participates in the physiological response to salinity change. *Mytilus edulis*, like many marine invertebrates, is an osmoconformer, unable to alter extracellular osmotic concentration in response to varying external salinity. Cell volume is regulated as a response to external salinity variations, at least in part, by alterations in the concentrations of free amino acids in the cytosol. The change in the composition and increase in the concentration of the free amino acid pool in response to increased environmental salinity is complex; the exact metabolic mechanisms for observed changes are not presently understood. Mobilization of cellular protein reserves is one possible mechanism for provisioning the cellular free amino acid pool (Bishop 1976).

Because of the correlation between allele frequencies at the *Lap* locus and environmental salinities, it is of special interest that total aminopeptidase-I activity acclimates to environmental salinity (Koehn 1978; Moore et al. 1980). Experimental alterations of salinity induce changes in both microdensitometrically and spectrophotometrically determined aminopeptidase-I activity, as an increase of enzyme activity with increased salinity and a decrease of enzyme activity with lowered salinity (Moore et al. 1980). The response of enzyme activity to changes in external salinity is virtually instantaneous in the lysosomal fraction (see below) and correlated with blood osmolality, but complete acclimination of enzyme activity requires about ten days (Koehn 1978). Salinity-induced enzyme activity changes are not due to changes in the *concentration* of aminopeptidase-I enzyme (Koehn unpublished).

The potential role of aminopeptidase-I in osmoregulation is further supported by known biochemical properties of the enzyme. Aminopeptidase-I is a dimer of size 68,000 daltons (Young et al. 1979) which has highest affinity for neutral and aromatic N-terminal residues of oligopeptides. The product of the catalyzed reaction is an amino acid. The enzyme is found in subcellular sites associated with active protein catabolism, having been immunocytochemically localized to both the intestinal brush border and lysosomes of digestive tubule cells of the digestive gland (Moore et al. 1980). Digestive tubule cells are rich in lysosomes (Owen 1972, Moore et al. 1978) and aminopeptidase-I has been demonstrated in cell-free lysosome preparations (Koehn unpublished). If the mobilization of cellular protein reserves is a mechanism for provisioning the cellular free amino acid pool during cell volume regulation, this will occur in lysosomes, since they are known to be involved in the turnover and catabolism of intercellular protein (Segal 1975; Dean 1977; Mortimore and Schworer 1977) and have relatively high concentrations of free amino acids in comparison to the cytosol (Tappel 1969; Ward and Mortimore 1978). Moore et al. (1980) have suggested that lysosomes in *M. edulis* may be involved in the degradation of intercellular proteins as a response to increased environmental salinities and that the activation of aminopeptidase-I (and other lysosomal enzymes) during this response represents an increase in the capability to degrade proteins and other compounds within the secondary lysosomes. Radioactive protein label is taken up by cell-free lysosomes, hydrolyzed to amino acids, and the equilibrium concentrations of protein and amino acids is dependent upon salinity of acclimation (Bayne et al., 1981).

The foregoing discussion provides a brief summary of the evidence that supports a relationship between variations of environmental salinity, aminopeptidase-I enzyme function, aminopeptidase-I allele frequency variations among natural populations experiencing differing salinities, and the potential role of the enzyme in the physiological mechanisms of osmoregulation. Nevertheless, additional information is required to bring these data to bear on the possible significance of aminopeptidase-I polymorphism. For example, what is the nature of the biochemical phenotypes among the aminopeptidase-I morphs upon which natural selection, in the form of environmental salinity, may act? What are the physiological consequences of these biochemical phenotypes? How might these consequences be different in different environments, and how do they ultimately result in fitness variation among individuals of different genotypes, if at all?

The answers to these questions are critical to our understanding of the aminopeptidase-I polymorphism and enzyme polymorphism at large. In the following sections, certain answers are formulated from the combined investigation of the biochemical properties of aminopeptidase-I genotypes, the dynamic behavior of a clinal boundary between genetically differentiated populations involving gene flow and mortality, and the physiological condition of animals whose dispersal influences the temporal and spatial instability of the cline.

Gene Flow and Mortality Among Populations

Samples of *Mytilus edulis* were randomly collected by hand during July (1976–78) in the intertidal zone at various sampling sites along the Northeastern tip of Long Island at the entrance to Long Island Sound from the Atlantic Ocean (detailed locality information is in Lassen and Turano 1978 and Koehn 1978). Shell length of individuals was measured to the nearest millimeter and data are presented on recruits (less than or equal to 15mm length) and resident adults (greater than 15mm shell length), since the annual settling cohort can be easily recognized in July as a group with shell size <15mm (Milkman and Koehn 1977). Settlement of pelagic larvae occurs in June–July. The frequencies of *Lap* alleles in the recruited cohort were usually identical to those in oceanic populations south of Cape Cod (Koehn et al. 1976), and therefore individuals of this cohort are termed "immigrants" to denote their origin from oceanic populations. The frequency of Lap^{94} in these oceanic populations varies closely around 0.55, while within Long Island Sound the frequency of this allele is approximately 0.15 (Koehn et al. 1976).

The decline in the frequency of Lap^{94} at the entrance to Long Island Sound occurs in less than 20 miles and the cline in immigrants and adults is very different (figure 5.1). In both 1976 and 1978, the immigrant cohort penetrated at least 12 and 17 miles, respectively, into Long Island Sound, whereas the cline in adults in both years was spatially coincident. Although the appearance of an immigrant cohort with high frequency of Lap^{94} is not as evident in 1977, in animals less than 10mm shell length the frequency of Lap^{94} was 0.40–0.55 at sites 1–4, suggesting that 1977 samples were taken later in the settlement period.

The annual immigration of oceanic larvae, carrying a high fre-

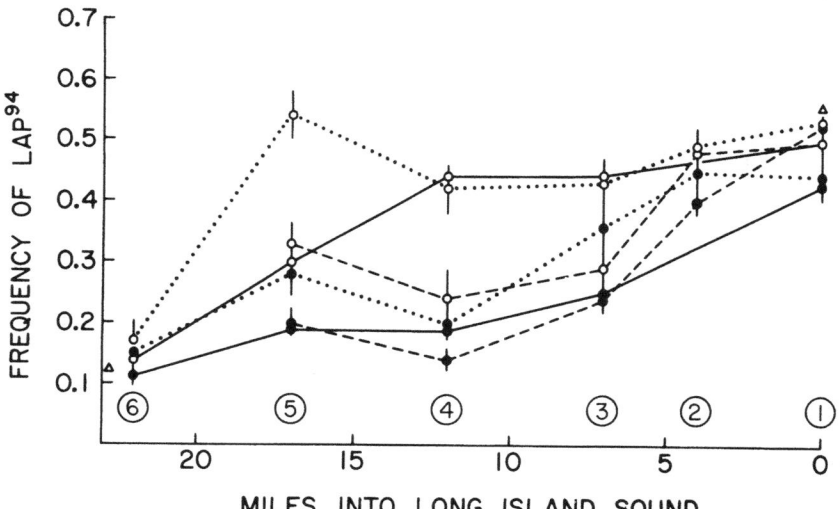

Figure 5.1. The clinal decrease in the frequency of Lap^{94} at the entrance to Long Island Sound for immigrants (o) and adults (•) in 1976 (———), 1977 (- - -), and 1978 (· · ·).

Note: The open triangles represent the average frequency of Lap^{94} in 20 replicate samples from Long Island Sound (left) and Atlantic Ocean (right). Data for 1976 are from Lassen and Turano (1978).

quency of the Lap^{94} allele, coupled with a low frequency of this allele at the same sites in resident adults, illustrates that annual gene flow by oceanically derived larvae is followed by mortality. When mussels at localities 1-5 are examined by 5mm size classes, the decrease in the frequency of Lap^{94} is constant at each site in 1977 and 1978, but different among sites (comparable data are not available for 1976). At localities 1 and 2, nearest the entrance to Long Island Sound, there is only a slight (statistically nonsignificant) decrease in frequency of Lap^{94} with size. The decrease of Lap^{94} with increase in shell size is most rapid at East Marion (locality 3) while there is a linear decrease in allele frequency with shell size at localities 4 and 5 (figure 5.2).

Figure 5.2 illustrates that the position of the allele frequency cline in immigrants varies from year to year, but that the postmortality cline in resident adults is spatially stable. This is further documented by the results of pairwise correlations among sites 1-5 between the frequency of the three common *Lap* alleles in recruits and adults in different years (table 5.1). For example, of the 27 pairwise tests of allele frequencies between years and allele frequencies with environmental salinity in immigrants, none gave significant correlations.

By way of contrast, of the 27 tests of adult allele frequencies, 14 are statistically significant. The behavior of the cline is most evident from an examination of Lap^{94}, which exhibits the greatest difference between oceanic and Long Island Sound populations. The spatial position of the immigrant cline varies from year to year, resulting in low correlations between years (−.351 to .662, table 5.1). The frequency of Lap^{94} in immigrants was not correlated with environmental salinity in any year, but the frequency of Lap^{94} in resident adults was highly correlated with salinity in each year and the frequencies among years at each site are relatively stable. Total aminopeptidase-I rapidly acclimates to an environmental salinity (Koehn 1978; Moore et al. 1980), and it is of particular interest that there is a significant correlation between measured enzyme activity and allele frequency in adults, but not immigrants (table 5.1). The variation in enzyme activity is a biologically relevant measure of environmental salinity variations.

We may conclude that the *Lap* cline is spatially and temporally

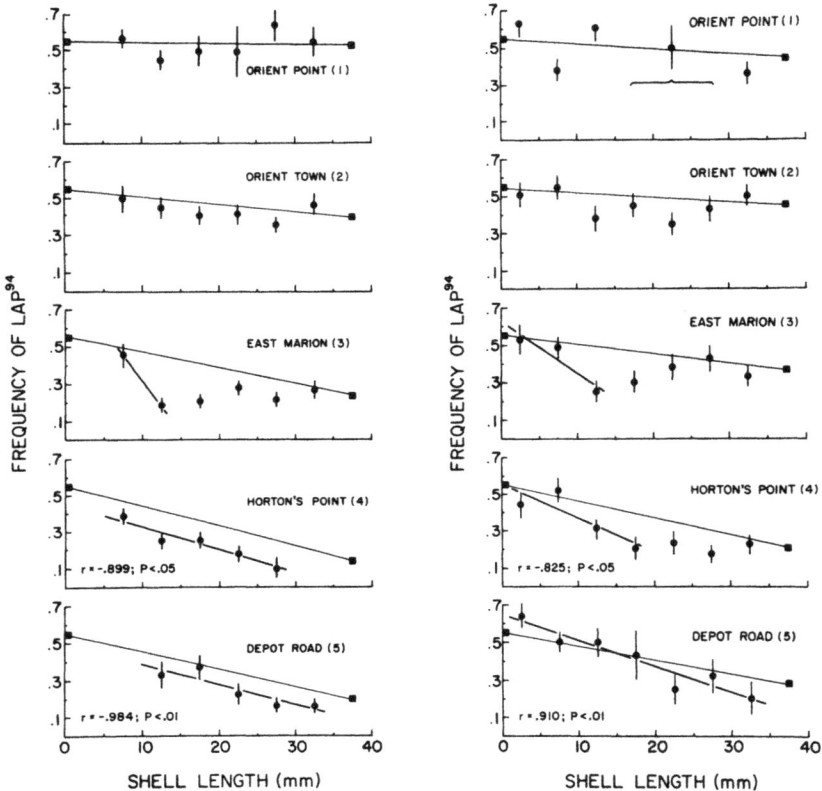

Figure 5.2. Size-dependent decrease in the frequency of *Lap⁹⁴* with increasing shell size at localities 1-5 (fig. 1) in 1977 (left) and 1978 (right).

Note: Where points are omitted, no individuals of that 5mm size-class were in the sample. Squares in each panel illustrate the differences in allele frequency between oceanic population (left) and average frequency of individuals > 15mm size at each site.

unstable in immigrants. The position must depend upon annual variations in the magnitude of gene flow, due to larval abundance, currents, and so forth. However, the cline is relatively stable in resident adults. The spatial position of the adult cline is exactly coincident with an environmental cline in salinity; despite annual variations in gene flow, mortality collapses the cline to the environmental boundary between Long Island Sound and the Atlantic Ocean.

Table 5.1. The Results of Correlation Tests Among Cline Samples (fig. 5.1) Between Lap Allele Frequency and Salinity and Between Allele Frequencies in Different Years in Recruits (Above Diagonal) and Adults (Below Diagonal).

	Enzyme Activity	1976	1977	1978	Salinity
Correlations of Lap^{98}					
Enzyme Activity		.805	−.425	.409	—
1976	−.816		−.498	.782	.846†
1977	−.663	.647		−.163	−.507
1978	−.948*	.648	.678		.424
Salinity	—	±.912	−.682	−.834	
Correlations of Lap^{96}					
Enzyme Activity		−.793	−.762	−.852	—
1976	−.867		.277	.601	−.892†
1977	−.965**	.956**		.538	−.663
1978	−.913*	.932*	.945*		−.725
Salinity	—	±.715	−.911*	−.894	
Correlations of Lap^{94}					
Enzyme Activity		.751	.800	.172	—
1976	.918**		.416	−.351	.873†
1977	.884*	.990**		.662	.750
1978	.949*	.900*	.909*		.049
Salinity	—	+.919*	.850	.849	

NOTE: The correlation between salinity and enzyme activity (in adults) was .964**. The center of each table, set off by dashed lines, gives the correlation between years for each allele in recruits and adults. df = 3, except where noted.
*$P < .05$ **$P < .01$ †df = 2

Gene Selection vs. Population Selection

The annual gene flow–mortality cycles described in the preceding section unequivocally demonstrate mortality of the immigrant cohort, but this does not provide critical evidence on the degree to which the dynamics of the cline are in any way directly a consequence of the *Lap* gene. As proposed by Levinton and Lassen (1978a, 1978b), the decrease in the frequency of the Lap^{94} allele in immigrants could be due to mortality of physiologically differentiated individuals arising from oceanic populations that are coincidentally marked by a high frequency of Lap^{94}. This "population selection" (figure 5.3) model assumes that the offspring of oceanic populations are physiologically adapted to oceanic environments, but not to the Long Island Sound environment. Oceanic larvae are passively dispersed into Long Island Sound, where the relative proportions of a mix between oceanic and Sound larvae determine the spatial position of the immigrant cline. The consequence of passive dispersal into Long Island

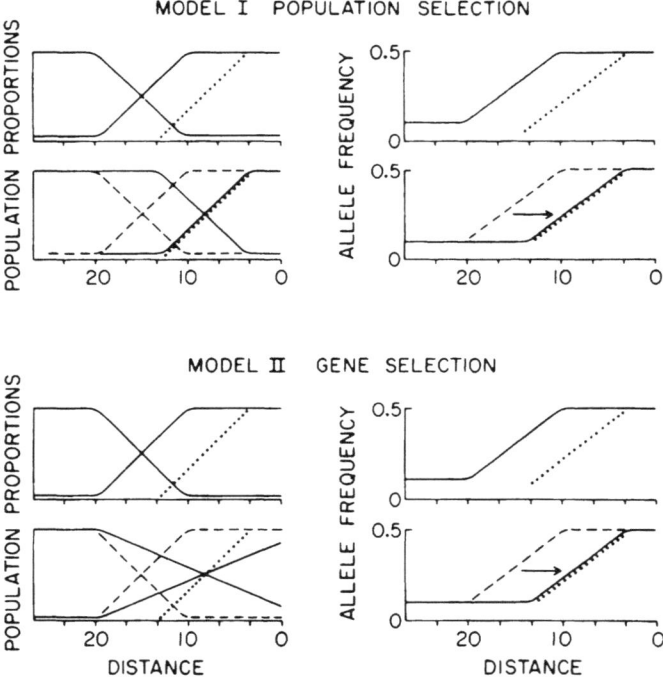

Figure 5.3. A comparison of population and gene selection models and the relationship of oceanic and Long Island Sound mixing proportions to spatial changes in the position of the Lap^{94} cline.

Note: In both models, immigration from oceanic populations initially displaces the cline to the left (upper panels of models I and II). In both models, mortality of immigrants shifts cline to the right (arrow), until it is coincident with the salinity gradient (dotted line). In model I, both the pre- and postmortality clines are due to population mixtures and the shift in cline position involves mortality of the immigrant cohort that is random with respect to Lap genotypes.
In model II, the premortality cline may be due to population mixing (or asymmetry in gene flow), but mortality is nonrandom with respect to Lap genotypes, so that allele frequency variation is not related simply to population proportions. See text. In all cases, dashed lines illustrate the premortality positions of immigrants and allele frequency after mortality.

Sound displaces the cline in a westerly direction (in this case to the left; figure 5.3). As settlement and growth proceeds, mortality of the immigrant cohort occurs because they are not physiologically able to adapt to the waters of Long Island Sound. Since this immigrant cohort is marked by a high frequency of Lap^{94}, the allele frequency cline is displaced by mortality to a position dictated by the environmental boundary (salinity) of Long Island Sound. This explanation does not require selection at the *Lap* locus. The cline is a consequence of population mixing followed by population mortality. Mortality in the immigrant cohort occurs randomly with respect to the *Lap* locus and the postmortality population consists primarily of offspring from Long Island Sound populations, characterized by a low frequency of Lap^{94}. The dynamic behavior of the cline can be explained merely by the differential success of populations that are coincidentally marked by differing frequencies of *Lap* alleles.

An alternative explanation requires different fitnesses among individuals of *Lap* genotypes. Mortality of the immigrant cohort is accompanied by differential mortality, which leads to a postselection population characterized by a low frequency of Lap^{94} genotypes, *specifically because of differences in aminopeptidase-I allozymes*. The two alternate models are contrasted in figure 5.3. Both models are consistent with the known characteristics of the *Lap* cline, but the underlying mechanisms are distinctly different in each.

A clear distinction between the two alternate models requires some direct evidence of differential mortality among *Lap* genotypes that can be directly related to aminopeptidase-I function (see below). However, there are several indirect lines of evidence that are informative and can also be used to distinguish between population and gene selection.

A model of population selection must be initiated by population mixing of oceanic and Long Island Sound larvae, since the postmortality population would consist of Sound populations (or their offspring) characterized by a low frequency of Lap^{94}. Several observations are inconsistent with population mixing. First, immigration onto the cline occurs by a massive settlement of individuals that produces a distinct "spike" in the size–frequency distribution. This cohort gradually merges with the overall size distribution during growth (Milkman and Koehn 1977). The frequency of Lap^{94} in this "spike" is typically the same as oceanic populations, suggesting that gene flow onto the cline is decidedly asymmetrical. Except for locality 6 (mile 22, figure 5.1), where Lap allele frequencies do not differ from other localities within Long Island Sound, no settling cohort has been observed in the region of the cline with a low frequency of Lap^{94}.

Population mixing gives certain expectations as to the correspondence of observed zygotic frequencies to Hardy-Weinberg proportions. For example, Lassen and Turano (1978) observed large heterozygote deficiencies near the midpoint of the allele frequency cline in the immigrant group and attributed this to population mixing, or the Wahlund effect. In their study, only a single heterozygote class was tested, since alleles were pooled. However, when multiple alleles are involved, population mixing produces different deviations from Hardy-Weinberg proportions in different heterozygotes, depending on the pattern of spatial variation of the individual alleles: alleles that positively co-vary among mixed populations exhibit an *excess* of heterozygotes relative to Hardy-Weinberg expectations, not a deficiency (Li 1969). In the case of the Lap cline, an excess of $Lap^{98\ 96}$ heterozygotes is expected, since there is positive covariance in the spatial variation of the Lap^{98} and Lap^{96} allele frequencies (table 5.2). Although the magnitude of deviation due to mixing will depend upon the population proportions that are mixed, decidedly different expectations arise for the individual Lap heterozygotes. If the immigrant cline is due to mixing of oceanic and Sound larvae, Wright's fixation index (F) for the $Lap^{98\ 96}$ heterozygote should be maximally negative near the midpoint of the cline while F should be maximally positive at the same point for the other two heterozygotes. In a comparison of the degree to which heterozygotes deviate from Hardy-Weinberg near the midpoint of the immigrant cline (table 5.2), it is apparent that observed deviations do not correspond to this expectation. F values computed for the $Lap^{98\ 96}$ heterozygote were not negative, as would be expected on the basis of population mixing, nor do deviations of the other two heterozygotes seem consistent with the expectations.

Table 5.2. A Comparison of Wright's Fixation Index, F, between Observed Values in 1976, 1977, and 1978 Near the Mid-Point of the Immigrant Cline (locality 5, figure 5.1) and Values Expected if the Immigrant Cline were Due to Larval Mixing Between Long Island Sound and Oceanic Populations.

Heterozygote	2σ	F_{exp}	F_{obs} 1976	1977	1978
$Lap^{98/96}$	.0458	−.211	.402	.320	.229
$Lap^{98/94}$	−.0987	.386	.403	.358	−.041
$Lap^{96/94}$	−.0854	.483	.687**	.460*	.178

NOTE: The mid-point of the immigrant cline would represent equal mixtures of contributions from populations in the Sound and Atlantic Ocean. With three alleles, an excess relative to Hardy-Weinberg expectations should occur in the heterozygote of those alleles between which there is positive covariance in spatial frequency variation, which in this case is $Lap^{98/96}$. The Hardy-Weinberg frequency of the $Lap^{98/96}$ heterozygote would be increased by twice the magnitude of the covariance between Lap^{98} and Lap^{96}, while the frequency of $Lap^{98/94}$ and $Lap^{96/94}$ would be decreased by twice the magnitude of the covariance between their constituent alleles (Li 1969).
*$P < .05$ **$P < .01$

If population mixing actually occurs, it may do so in a pattern unlike that supposed in the computations in table 5.2. Nevertheless, the population selection model requires random mortality with respect to the *Lap* locus in the immigrant population, so that the distribution and magnitudes of F will be virtually identical between the immigrants and resident adults when compared among localities (i.e., the cline is shifted, but mortality is *random* with respect to *Lap* genotypes). In addition, significant values of F are expected only at those localities where intermediate allele frequencies occur (i.e., where there is mixing). An examination of tables 5.3–5 demonstrates that neither of these expectations is realized. Significant heterozygote deficiencies are observed at locality 6 in all three years, a locality at which there is no evidence of mixing; *Lap* allele frequencies at this site are identical to other populations in Long Island Sound. Significant heterozygote deficiencies were observed in resident adults in 1976 and 1977 at locality 1, a site where allele frequencies do not differ from other oceanic populations. Last, a comparison of immigrants and resi-

Table 5.3. Deviations of Observed Aminopeptidase-I Heterozygotes from Hardy-Weinberg Expectations in Immigrants and Residents in 1976.

Locality	Immigrants				Residents			
	N	98/96	96/94	98/94	N	98/96	96/94	98/94
1	—	—	—	—	236	-.060	.132*	.056
1a	80	.190**	.216**	-.012	358	.151**	-.084	.163**
3	52	.025	.009	.139	386	.095	.327**	.378**
4	74	.294**	.381**	.132	151	.000	.493**	.302**
5	23	.402	.687**	.403	239	.342**	.356**	.280**
6	72	.268**	.214*	.289**	331	.130*	.470**	-.076

NOTE: (data from Lassen and Turano 1978). Deviations are given as Wright's fixation index, F, estimated as $1 - H_o/H_e$ and tested as $\chi^2 = NF^2 (k - 1)$, where H_o is the number of observed heterozygotes, H_e is the number of heterozygotes expected by Hardy-Weinberg and k is the number of alleles. Degrees of freedom = $k(k - 1)/2$ (Li and Horovitz, 1953). Localities are in figure 5.1.
*$P < .05$ **$P < .01$

Table 5.4. Data from 1977 in Same Form as Table 5.3.

Locality	Immigrants				Residents			
	N	98/96	96/94	98/94	N	98/96	96/94	98/94
1	118	.368**	.015	-.075	151	.082	-.064	.028
2	52	.029	-.203	-.019	226	-.041	.124	.178**
3	58	.309*	-.194	.188	248	-.056	.387**	.333**
4	34	.268	.452**	-.214	104	.228*	.136	.210*
5	26	.320	.460*	.358	158	.101	.121	.226*
6	78	.198	.437**	.360**	89	.290**	.480**	.520**

*$P < .05$ **$P < .01$

Table 5.5. Data from 1978 in Same Form as Table 5.3.

Locality	Immigrants				Residents			
	N	98/96	96/94	98/94	N	98/96	96/94	98/94
1	86	.126	-.015	.211	70	-.111	.122	.247*
2	80	-.235*	.015	.136	137	-.116	.246**	.195*
3	75	-.154	-.082	.141	122	.146	.437**	.299**
4	80	.181	-.111	.284*	110	.169	.248**	.600**
5	74	.229	.178	-.041	73	.124	.751**	-.264**
6	81	.252*	.077	.400**	78	-.166	.470**	-.376**

*$P < .05$ **$P < .01$ df = 3

dents in 1978 (table 5.5) shows that the distribution and magnitudes of F are very different between the two groups.

In summary, observed zygotic frequencies do not seem consistent with a simple model of population mixing, followed by random mortality at the *Lap* locus. The magnitudes of F in the postmortality cline are both larger, often of opposite sign, and distributed differently than would be expected by the Wahlund effect.

Biochemical Phenotypes of the Aminopeptidase-I Polymorphism

A clear distinction between the two alternate models in figure 5.3 requires some direct evidence of differential mortality among *Lap* genotypes that can be directly related to aminopeptidase-I enzyme function. This depends in part on known differences in biochemical function among genotypes. A detailed analysis of kinetic parameters of individual aminopeptidase-I phenotypes has been done. As the results have reported in detail elsewhere (Koehn and Siebenaller 1981), they are only briefly summarized here. There were no significant among-genotype differences in: (1) apparent K_m of L-leucyl-4-methoxy-β-naphthylamide over the temperature range 5–30°C, (2) apparent K_m of phenylalanine-glycine over the temperature range 5–25°C, (3) sensitivity of V_{max} and K_m of substrate to pH, tested over the range pH 6.5–9.0, (4) inhibition by amino acids, and (5) heat denaturation. Arrhenius activation energies (E_a) of the six genotypes were homogeneous, indicating the activation enthalpies ($\Delta H^{\ddagger}$; $\Delta H^{\ddagger} = E_a - RT$) were identical. In summary, of the many kinetic parameters of the aminopeptidase-I enzyme that we have examined, none appeared to differ among genotypes.

There are significant differences in specific activities among *Lap* genotypes in natural populations (Koehn 1978; table 5.6). Genotypes involving the Lap^{94} allele exhibit the highest specific activities at Shinnecock, an oceanic locality on the south shore of Long Island. However, in the Stony Brook population, exposed to approximately 23 ppt environmental salinity in Long Island Sound, there are also differences in specific activities among *Lap* genotypes, but the $Lap^{94\ 94}$ homozygote exhibits lowest relative specific activity (table 5.6). Hence, the lower salinity of Long Island Sound has several consequences. First, there is a reduction in overall levels of aminopeptidase-I activity (Koehn 1978). Second, there is a substantial decrease in the frequency of Lap^{94}, which occurs at the entrance to Long Island Sound coincidental with the diminution of salinity. Third, the Lap^{94} genotypes exhibit highest specific activity in oceanic salinity, but the $Lap^{94\ 94}$ homozygote exhibits lowest specific activity in the reduced salinity of Long Island Sound. The specific activity of this genotype exhibits inverse compensation, being highest at high salinity and lowest in low salinity populations relative to other genotypes.

When enzyme is electrophoresed in gels containing anti-aminopeptidase-I antibody, the heights of the immunoprecipitate "rockets" are proportional to the number of aminopeptidase-I molecules (Laurell 1966). While genotypes of the Lap^{94} allele exhibit relatively greater levels of specific activity (units of activity per milligram total protein) at Shinnecock, there are no differences in the concentration of aminopeptidase-I enzyme among genotypes (table 5.7). In the Stony Brook population, the low relative specific activity of the $Lap^{94\ 94}$ homozygote (table 5.6) is due to the

Table 5.6. Analysis of Rank Genotype Means in Specific Activity of Aminopeptidase-I in Homogenates of Digestive Gland from Shinnecock (oceanic salinity) and Stony Brook (estuarine salinity).

Shinnecock		Stony Brook	
Genotype	Group Sums	Genotype	Group Sums
$Lap^{94\ 94}$	19.0	$Lap^{96\ 94}$	14.0
$Lap^{96\ 94}$	25.0	$Lap^{98\ 94}$	29.0
$Lap^{98\ 94}$	29.0	$Lap^{98\ 96}$	31.0
$Lap^{98\ 96}$	35.0	$Lap^{96\ 96}$	33.0
$Lap^{96\ 96}$	39.0	$Lap^{98\ 98}$	41.0
$Lap^{98\ 98}$	42.0	$Lap^{94\ 94}$	
FS = 12.17; P < .05		FS = 15.73; P < .01	

NOTE: FS = Friedman's randomized block statistic. Analysis was on ten samples from each locality. From Koehn et al. (1980a).

Table 5.7. One-Way Analysis of Variance of the Average Laurell Rocket Heights/mg Total Protein of *Lap* Genotypes from Shinnecock (Oceanic Salinity) and Stony Brook (Estuarine Salinity).

	98/98	98/96	96/96	98/94	96/94	94/94	$F_{5,24}$
Shinnecock	16.33	15.84	17.55	18.83	17.63	16.98	.98; ns
Stony Brook	*18.85*	*17.76*	20.61	*17.38*	*17.89*	15.76	3.63; P < .025

NOTE: Genotypes at Stony Brook (in italics) constitute a homogeneous group by the simultaneous test procedure (Sokal and Rohlf 1969). From Koehn et al. (1980a).

lower concentration of aminopeptidase-I enzyme in the genotype, since all other genotypes constitute a homogeneous group (table 5.7). Despite the lower concentration of enzyme in this homozygote at Stony Brook, all genotypes of the Lap^{94} allele exhibit highest levels of specific activity per quantity of aminopeptidase-I enzyme. Genotypes without the Lap^{94} allele exhibit only about 83 percent of the catalytic efficiency of Lap^{94} genotypes when standardized by enzyme concentration (table 5.8). Hence, genotypes containing Lap^{94} exhibit greater catalytic efficiency (K_{cat}) than non-Lap^{94} genotypes. This higher catalytic efficiency produces higher specific activities in natural populations, though the $Lap^{94\;94}$ homozygote is lower in areas of reduced salinity because of a reduction in the quantity of aminopeptidase-I enzyme.

In summary, the cline in frequency of Lap^{94} at the eastern entrance of Long Island Sound represents a decrease in frequency of an allele with high catalytic efficiency between high salinity–high enzyme activity populations and low salinity–low enzyme activity populations and the enzyme concentration of the homozygote

Table 5.8. One-Way Analysis of Variance of the Average Specific Activities/Rocket Height for *Lap* Genotypes With and Without the *Lap*^{94} Allele in Samples from Shinnecock (Oceanic Salinity) and Stony Brook (Estuarine Salinity).

	Shinnecock	Stony Brook
Lap^{94} genotypes	5,431	5,270
non-Lap^{94} genotypes	4,300	4,590
	$F_{1,28}$ = 4.40; P < .05	$F_{1,30}$ = 5.44; P < .05

SOURCE: Koehn et al (1980a).

of this allele also declines. The decreased concentration of enzyme in this homozygote could be due to instability of the enzyme-membrane interaction, but this has not been studied.

Physiological Consequences of Aminopeptidase-I Polymorphism

Although the foregoing paragraphs have concentrated on the change in environmental salinity at the entrance to Long Island Sound and the effect that this environmental change may have on individual *Lap* genotypes, any hypothesis that attempts to explain differentiation of *Lap* allele frequencies among different areas along the Atlantic coast must take account of the effect of both salinity and temperature upon both digestive and osmoregulatory physiologies. While the frequency of Lap^{94} declines in estuaries south of Cape Cod, it also declines in a northward direction at Cape Cod, where there is no change in environmental salinity, but a decrease in water temperature (Koehn et al. 1976). Is the high catalytic efficiency of Lap^{94} sufficient to explain the higher frequency of this allele in normal oceanic salinity habitats at lower latitudes and its low frequency elsewhere? If the answer to this question is affirmative, there must be both a conceptual and detectable link among biochemical variation, physiological variation, and allele frequency change.

High temperatures elicit increased enzyme protein concentration in digestive enzymes (Hazel and Prosser 1974). High salinity induces high aminopeptidase-I activity as an apparent adaptive response to effective cell volume regulation. We would therefore expect the evolution of a high catalytic rate allozyme under high temperature, oceanic salinity conditions. At warm temperatures and oceanic salinities, increased reaction rate is adaptive, because effective osmoregulation requires relatively high enzyme activity and elevated activity is thermally compensatory. Lap^{94} is found in high frequency only where these two environmental conditions occur together (Koehn et al. 1976). Lap^{94} phenotypes seem to extend the thermal compensatory capacity of the aminopeptidase-I enzyme. However, at lower salinities or lower temperatures, high catalytic efficiency could be a disadvantage as a function of the potential "cost" of increased V_{max} in digestive and/or osmoregulatory physiologies.

To let the absorption of the products of protein digestion proceed too rapidly would be a positive disadvantage (Cornish-Bowden 1976). This would include both ingested protein and cellular protein reserves at times when environmental nitrogen is limiting. We might then suggest that at low salinities, where overall aminopeptidase-I activity is low, Lap^{94} phenotypes will have a decreased efficiency of nitrogen assimilation, because of the imbalance between uptake rate of amino acids from either the gut or cellular protein reserves and the rate of amino acid utilization. At times when environmental nitrogen is limiting, we would expect individuals with the Lap^{94} allele to more rapidly exhaust protein reserves. Individuals with this high catalytic efficiency allele metabolize protein to cellular amino acids at a higher rate (Hilbish et al. 1982).

We have tested this hypothesis (Koehn et al. 1980b) in our examination of an index of physiological condition that would reflect genotypic differences in energetic stress and thereby provide an explanation for differential mortality among *Lap* genotypes in the immigrant cohort of the cline at the entrance to Long Island Sound. The dry tissue weight of molluscs, expressed in relation to some shell parameter (e.g., weight, volume, or size) has frequently been used as a physiological condition index (Baird 1966; Lent 1967). In *Mytilus edulis*, this index varies seasonally (Dare and Edwards 1975) as the weight of germinal tissue varies (due to the buildup and liberation of gametes) and the weight of somatic tissue varies, as nutrient reserves are stored and utilized (Gabbott and Stephenson 1974; Gabbott 1976). Locality 3 on the cline (fig. 5.1) was selected for this study because of the apparent rapid decrease in the frequency of Lap^{94} with increase in shell size (fig. 5.2). The relationship between dry tissue weight and shell length was compared between groups of individuals with and without the Lap^{94} allele in both July (immediately following settlement) and October 1979. In July, a time of high food availability, there were no significant differences between groups in the exponents or intercepts relating dry tissue weight to shell length. In October, a period when nutrient reserves are declining, there was no difference between groups in exponents, but intercepts differed significantly between groups (table 5.9). Thus, animals possessing the Lap^{94} allele, in either heterozygous or homozygous condition, lost a significantly larger amount of tissue (ca. 15%) than animals without the Lap^{94} allele. This loss occurred during the period July–October, when mortality produced a decrease in the frequency of Lap^{94} allele on the cline. Since intercepts (but not exponents) differed, there was a proportional weight loss in animals in all sizes.

Conclusion and Comments

We are encouraged by the ability to predict certain physiological consequences of biochemical differences among genotypes. There are, nevertheless, numerous points that must be investigated before the interpretations in the foregoing paragraphs can be firmly established. For example, more information is needed on the fate of nitrogen in the metabolism of individuals on the cline and how this may differ among individuals of differing genotype. Detailed measurements of nitrogen availability are needed as well as estimates of rates of nitrogen turnover in individuals of known genotype. Recent studies by Livingston et al. (1979) have suggested

Table 5.9. Analysis of Covariance of the Regressions of $\log_{10}$ Dry Tissue Weight on $\log_{10}$ Shell Length in Mussels With and Without the Lap^{94} Allele Sampled from Locality 3 (fig. 5.1) in October 1979.

Source of Variation	d.f.	ss	ms	F
Total Residual	401	5.114	0.013	
Difference in Slopes	1	0.033	0.033	2.625; ns
Subtotal	402	5.148	0.013	
Difference in Intercepts	1	0.518	0.518	40.465; $P \ll .001$
Total	403	5.666		

that there is rapid turnover of amino-nitrogen in animals exposed to fluctuating salinities. The decreased physiological condition of all animals in October (fig. 5.4) could reflect the increased "osmotic work" that is required at the environmental boundary between Long Island Sound and the Atlantic Ocean. More detailed information on the magnitude and pattern of salinity variation in the environs of the cline is necessary so that interpretations may be experimentally investigated under controlled conditions.

While numerous questions remain unanswered, it is possible to make several points. Larval dispersal in *Mytilus edulis* lasts from three to seven weeks (Bayne 1976), but despite this, significant

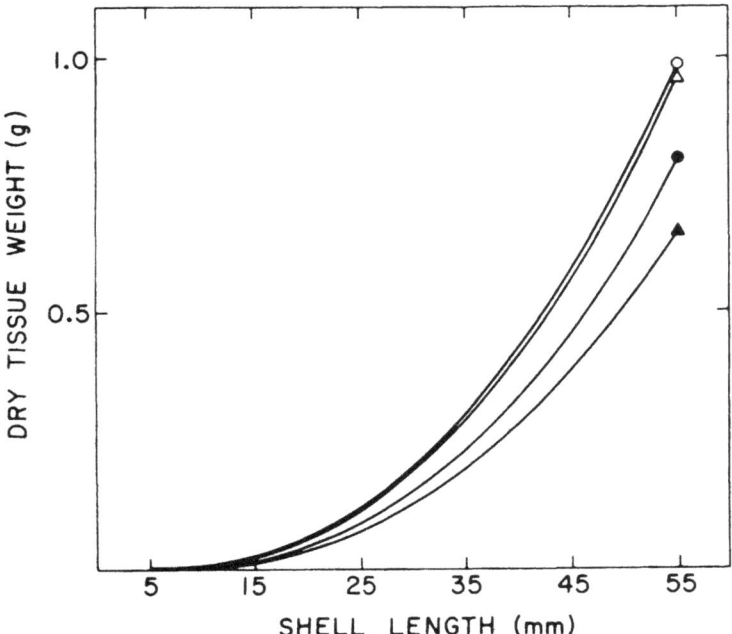

Figure 5.4. The relationship between dry tissue weight and shell length in July 1979 (○, △) and October 1979 (●, ▲) for individuals with (△, ▲) and without (○, ●) the *Lap*^{M} allele.

Note: Statistical analyses were performed on $\log_{10}$ transformed variables, but the illustrated curves are the computed regression lines, detransformed. Each regression involved > 200 points. From Koehn et al. (1980b).

genetic differences can occur between populations separated by very little distance. Environmental differences that affect population differentiation are relatively small; the salinity of Long Island Sound is only about 8 ppt lower than that of the Atlantic Ocean, and there are only small differences in annual temperature variation between oceanic waters on each side of Cape Cod. Nevertheless, these environmental differences are sometimes sufficient to produce substantial mortality which counterbalances gene flow. Biochemical and physiological data demonstrate that this mortality can be at least partially attributed to differences in the way aminopeptidase-I phenotypes adapt animals to different environments. While it would be naïve to suppose that this is the only gene to which natural selection is directed in these cases, the data emphasize that certain environmental factors can have detectable evolutionary influences on individual genes.

Although our observations bear upon the general question of the adaptive nature of enzyme polymorphism that has so dominated evolutionary genetics for the past decade, it is difficult to justify generalizing these results to other enzymes in other species. The aminopeptidase-I enzyme in *Mytilus edulis* is not embedded in a complex metabolic pathway, but catalyzes the degradation of oligopeptide substrates that enter the animal from the environment or are mobilized from cellular protein reserves. As a consequence, the *Lap* gene may be more specifically targeted by natural selection than genes involved in, for example, intermediary metabolism. It would, however, be erroneous to imply that aminopeptidase-I is the only enzyme that is important in osmoregulation and temperature adaptation. Indeed, the activities of other lysosomal enzymes are salinity induced, much like aminopeptidase-I (Moore et al. 1980). It is possible that other enzymes, physiologically integrated with aminopeptidase-I, could also exhibit adaptively important biochemical variations related to salinity and temperature variation. The investigation of biochemically and physiologically integrated enzymes should be a more informative approach to understanding genetic variation in natural populations than studies involving randomly selected and biochemically diverse genes.

ACKNOWLEDGMENTS

This study could not have been possible without the active collaboration of many persons. Each brought to this project considerable expertise and imagination and it is impossible to disentangle many of their contributions from my own. Major contributions were made by Drs. Brian L. Bayne and Michael N. Moore of the Institute of Marine Environmental Research, Plymouth, England. Much of the work described in this paper was done collaboratively with them. Both have served as a source of instruction and guidance to me and I value their warm friendship and active scientific collaboration. Dr. Roger I. E. Newell, while serving as a NATO postdoctoral fellow, contributed significantly to the physiological aspects of this study and I am especially grateful for the expert technical assistance of Christine Newell. Both have been valuable colleagues during the course of this study. I have benefitted from constructive discussions with Drs. J. F. Siebenaller and N. Arnheim. I am grateful for the laboratory and field assistance of Peter Bellacera, Frank DePew, John Hall, Jerry Hilbish, Fred Immerman, David Innes, and Anthony Zera. Ms. Patricia Haines typed the manuscript. This research was supported by grants DEB 7706074 and DEB 7908862 from the National Science Foundation, USPHS grant GM 21133, and NATO grant 1440. This is contribution number 352 from the Program in Ecology and Evolution, State University of New York, Stony Brook, New York 11794.

REFERENCES

Ahmad, M., D. O. S. Skibinski, and J. A. Beardmore. 1977. An estimate of the amount of genetic variation in the common mussel, *Mytilus edulis. Biochem. Genet.* 15:833–846.

Ayala, F. J., D. Hedgecock, G. S. Zumwalt, and J. W. Valentine. 1973. Genetic variation in *Tridacna maxima:* An ecological analog of some unsuccessful evolutionary lineages. *Evol.* 27:177–191.

Baird, R. H. 1966. Factors affecting the growth and condition of mussels (*Mytilus edulis*). *Fish. Invest. Minist. Agric. Fish. Food* (G. B.) Serv. II 25:1–33.

Bayne, B. L. 1976. The biology of mussel larvae. In B. L. Bayne, ed., *Marine Mussels: Their Ecology and Physiology*, pp. 81–120. Cambridge; Cambridge University Press.

Bayne, B. L., N. M. Moore and R. K. Koehn. 1981. Lysosomes and the response by *Mytilus edulis* L. to an increase in salinity. *Mar. Biol. Lettrs.* 2:193–204.

Bijlsma, R. and C. van der Meulen-Bruijns. 1979. Polymorphism of G6pd and 6Pgd loci in *Drosophila melanogaster*. III. Developmental and biochemical aspects. *Biochem. Genet.* 17:1131–1144.

Bijlsma-Meeles, E. and W. van Delden. 1974. Intra- and interpopulation selection concerning the alcohol dehydrogenase locus in *Drosophila melanogaster*. *Nature.* 247:369–371.

Bishop, S. H. 1976. Nitrogen metabolism and excretion: Regulation of intracellular amino acid concentration. In M. Wiley, ed., *Estuarine Processes*, 1:414–431. New York: Academic Press.

Buroker, N. E., W. K. Hershberger, and K. K. Chew. 1975. Genetic variation in the Pacific Oyster, *Crassostrea gigas*. *J. Fish. Research Board of Canada* 32:2471–2477.

Clarke, B. 1975. The contribution of ecological genetics to evolutionary theory: Detecting the direct effects of natural selection on particular polymorphic loci. *Genetics.* 79s:101–113.

Cornish-Bowden, A. 1976. The effect of natural selection on enzymatic catalysis. *J. Mol. Biol.* 101:1–9.

Dare, P. J. and D. B. Edwards. 1975. Seasonal changes in flesh weight and biochemical composition of mussels (*Mytilus edulis* L.) in the Conway estuary, North Wales. *J. Exp. Mar. Biol. Ecol.* 18:89–97.

David, J. R., C. Bocquet, M.-F. Arens, and P. Fouillet. 1976. Biological role of alcohol dehydrogenase in the tolerance of *Drosophila melanogaster* to aliphatic alcohols: utilization of an ADH-null mutant. *Biochem. Genet.* 14:989–998.

Day, T. H., P. C. Hillier, and B. Clarke. 1974. The relative quantities and catalytic activities of enzymes produced by alleles at the alcohol dehydrogenase locus in *Drosophila melanogaster*. *Biochem. Genet.* 11:155–166.

Dean, R. T. 1977. *Lysosomes*. London: Edward Arnold.

Endler, J. A. 1977. *Geographic Variation, Speciation, and Clines*. Princeton: Princeton Univ. Press.

Fucci, L., L. Gaudio, R. Rao, A. Spanò, and M. Carfagna. 1979. Prop-

erties of the two common electrophoretic variants of phosphoglucomutase in *Drosophila melanogaster*. *Biochem. Genet.* 17:825–836.
Gabbott, P. A. 1976. Energy Metabolism. In B. L. Bayne, ed. *Marine Mussels: Their Ecology and Physiology*, pp. 293–355. Cambridge: Cambridge University Press.
Gabbott, P. A. and R. R. Stephenson. 1974. A note on the relationship between the dry weight condition index and the glycogen content of adult oysters (*Ostrea edulis* L.) kept in the laboratory. *J. Cons., Cons. Int. Expl. Mer.* 35:359–361.
Gibson, J. B. 1970. Enzyme flexibility in *Drosophila melanogaster*. *Nature* 227:959.
Hazel, J. R. and C. L. Prosser. 1974. Molecular mechanisms of temperature compensation in poikilotherms. *Physiol. Rev.* 54:620–677.
Hedrick, P. W., M. R. Ginevan, and E. P. Ewing. 1976. Genetic polymorphism in heterogeneous environments. *Ann. Rev. Ecol. System.* 7:1–32.
Hilbish, T. J., L. E. Deaton and R. K. Koehn. 1982. Effect of an allozyme polymorphism on regulation of cell volume. *Nature* 298:688–689.
Koehn, R. K. 1969. Esterase heterogeneity: Dynamics of a polymorphism, *Science*. 163:943–944.
Koehn, R. K. 1978. Physiology and biochemistry of enzyme variation: The interface of ecology and population genetics. In P. F. Brussard, ed. *Ecological Genetics: The Interface*, pp. 51–72. New York: Springer-Verlag.
Koehn, R. K., B. L. Bayne, M. N. Moore, and J. F. Siebenaller. 1980a. Salinity related physiological and genetic differences between populations of *Mytilus edulis*. *Biol. J. Linn. Soc.* 14:319–334.
Koehn, R. K., R. Milkman, and J. B. Mitton. 1976. Population genetics of marine pelecypods. IV. Selection, migration and genetic differentiation in the Blue Mussel, *Mytilus edulis*. *Evol.* 30:2–32.
Koehn, R. K. and J. B. Mitton. 1972. Population genetics of marine pelecypods. I. Ecological heterogeneity and evolutionary strategy at an enzyme locus. *Am. Nat.* 106:47–56.
Koehn, R. K., R. I. E. Newell, and F. Immermann. 1980b. Maintenance of an aminopeptidase allele frequency cline by natural selection. *Proc. Natl. Acad. Sci., U.S.* 77:5385–5389.
Koehn, R. K. and J. F. Siebenaller. 1981. Biochemical studies of aminopeptidase-I polymorphism in *Mytilus edulis*. II. Dependence of reaction rate on physical factors and enzyme concentration. *Biochem. Genet.* 11/12:1143–1162.

Lassen, H. H. and F. J. Turano. 1978. Clinical variation and heterozygote deficit at the Lap-locus in *Mytilus edulis*. *Mar. Biol.* 49:245-254.

Laurell, C.-B. 1966. Quantitative estimation of proteins by electrophoresis in agarose gels containing antibodies. *Anal. Biochem.* 15:45-52.

Lent, C. M. 1967. Effect of habitat on growth indices in the Ribbed Mussel, *Modiolus (Arcuatula) demissus*. *Chesapeake Sci.* 8:221-227.

Levinton, J. S. 1975. Levels of genetic polymorphism at two enzyme encoding loci in eight species of the genus *Macoma* (Mollusca: Bivalvia). *Mar. Biol.* 33:41-47.

Levinton, J. S. and D. L. Fundiller. 1975. An ecological and physiological approach to the study of biochemical polymorphisms. *Proc. 9th Europ. Mar. Biol. Symp.*, pp. 165-178.

Levinton, J. S. and R. K. Koehn. 1976. Population genetics of mussels. In B. L. Bayne, ed. *Marine Mussels: Their Ecology and Physiology*, pp. 357-384, Cambridge: Cambridge University Press.

Levinton, J. S. and H. H. Lassen. 1978a. Selection, ecology, and evolutionary adjustment within bivalve mollusc populations. *Phil. Trans. R. Soc. Lond. B.* 284:403-415.

Levinton, J. S. and H. H. Lassen. 1978b. Experimental mortality studies and adaptation at the Lap locus in *Mytilus edulis*. In B. Battaglia and J. A. Beardmore, eds. *Marine Organisms*, pp. 229-254. New York: Plenum.

Levinton, J. S. and T. H. Suchanek. 1978. Geographic variation, niche breadth and genetic differentiation at different geographic scales in the mussels *Mytilus californianus* and *M. edulis*. *Mar. Biol.* 49:363-375.

Lewontin, R. C. 1974. *The Genetic Basis of Evolutionary Change*. New York: Columbia University Press.

Li, C. C. 1969. Population subdivision with respect to multiple alleles. *Ann. Human Genet.* 33:23-29.

Li, C. C. and D. G. Horovitz. 1953. Some methods of estimating the inbreeding coefficient. *Am. J. Human Genet.* 5:107-117.

Livingston, D. R., J. Widdows, and P. Fieth. 1979. Aspects of nitrogen metabolism of the common mussel, *Mytilus edulis:* adaptation to abrupt and fluctuating changes in salinity. *Mar. Biol.* 53:41-55.

McDonald, J. and J. Avise. 1976. Evidence for the adaptive significance of enzyme activity levels: interspecific variation in αGPDH and ADH in *Drosophila*. *Biochem. Genet.* 14:347-355.

McDonald, J. F. and F. J. Ayala. 1978. Genetic and biochemical basis of enzyme activity variation in natural populations. I. Alcohol dehydrogenase in *Drosophila melanogaster*. *Genetics*. 89:371-388.

McDonald, J. F., G. K. Chambers, J. David, and F. J. Ayala. 1977. Adaptive response due to changes in gene regulation: a study with Drosophila. *Proc. Natl. Acad. Sci., U.S.* 74:4562-4566.

Merritt, R. B. 1972. Geographic distribution and enzymatic properties of lactate dehydrogenase allozymes in the Fathead Minnow, *Pimaphales promelas*. *Am. Nat.* 196:173-184.

Milkman, R. and R. K. Koehn. 1977. Temporal variation in the relationship between size, numbers, and an allele-frequency in a population of *Mytilus edulis*. *Evol.* 31:103-115.

Moore, M. N., R. K. Koehn, and B. L. Bayne. 1980. Leucine aminopeptidase (aminopeptidase-I), N-acetyl-β-hexoseaminidase and lysosomes in the mussel, *Mytilus edulis* L., in response to salinity changes. *J. Exp. Zool.* 214:239-249.

Mortimore, G. E. and C. M. Schworer. 1977. Induction of autophagy by amino-acid deprivation in perfused rat liver. *Nature* 270:174-176.

Murdock, E. A., A. Ferguson, and R. Seed. 1975. Geographic variation in leucine aminopeptidase in *Mytilus edulis* L. from the Irish coasts. *J. Exp. Mar. Biol. Ecol.* 19:33-41.

Owen, G. 1972. Lysosomes, peroxisomes and bivalves. *Sci. Progr., Oxford.* 60:299-318.

Place, A. R. and D. A. Powers. 1979. Genetic variation and relative catalytic efficiencies: lactate dehydrogenase B allozymes of *Fundulus heteroclitus*. *Proc. Natl. Acad. Sci., U.S.* 76:2354-2358.

Rasmuson, B., L. R. Nilson, M. Rasmuson, and E. Zeppezauer. 1966. Effects of heterozygosity on alcohol dehydrogenase (ADH) activity in *Drosophila melanogaster*. *Hereditas* 56:313-316.

Schaal, B. A. and W. W. Anderson. 1974. An outline of techniques for starch gel electrophoresis of enzymes from the American Oyster *Crassostrea virginica* Gmelin. Georgia Marine Science Center Technical Report Series Number 74-3.

Segal, H. L. 1975. Lysosomes and intracellular protein turnover. In J. T. Dingle, and R. T. Dean, eds. *Lysosomes in Pathology and Biology* 4:295-302. Amsterdam: Elsevier/North-Holland.

Sokal, R. R. and F. J. Rohlf. 1969. *Biometry*. San Francisco: W. H. Freeman and Co.

Tappel, A. L. 1969. Lysosomal enzymes and other components. In *Lysosomes in Biology and Pathology*. J. T. Dingle, and H. B. Fell, eds. 2:207-244. Amsterdam: Elsevier/North-Holland.

Theisen, B. F. 1978. Allozyme clines and evidence of strong selection in three loci in *Mytilus edulis* L. (Bivalvia) from Danish waters. *Ophelia* 17:135-142.

Vigue, C. L. and F. M. Johnson. 1973. Isozyme variability in species of the genus Drosophila. VI. Frequency-property-environment relationships of allelic alcohol dehydrogenases in *D. melanogaster*. *Biochem. Genet.* 9:213-227.

Ward, R. and P. Hebert. 1972. Variability of alcohol dehydrogenase activity in a natural population of *Drosophila melanogaster*. *Nature New Biol.* 236:243.

Ward, W. F. and G. E. Mortimore. 1978. Compartmentation of intracellular amino acids in rat liver. Evidence for an intralysosomal pool derived from protein degradation. *J. Biol. Chem.* 253:3581-3587.

Young, J. P. W., R. K. Koehn, and N. Arnheim. 1979. Biochemical charactaerizations of "Lap," a polymorphic aminopeptidase from the Blue Mussel, *Mytilus edulis*. *Biochem. Genet.* 17:305-323.

6 Genetic Variation, Environmental Heterogeneity, and Evolutionary Stability

Peter S. Dawson and Russel A. Riddle

One of the areas of population biology in which meaningful attempts at a synthesis between genetic and ecological approaches have been made involves the effects of environmental heterogeneity on the genetical structure of populations. Beginning with the seminal papers of Levene (1953) and Dempster (1955), a rich theoretical literature has developed concerning the maintenance of polymorphism in complex environments.

Curiously, this topic received rather little attention at the Syracuse Symposium on Population Biology (Lewontin 1968). At that time, we had just begun to learn of the rather startlingly large amounts of genetic variation present in natural populations from the pioneering work of Lewontin and Hubby (1966) and Harris (1966). Richard Levins gave a short presentation of his fitness set approach to adaptation of populations to an environment consisting of two patch types (Levins 1968a). C. H. Waddington reminded us that environmental heterogeneity must necessarily affect fitness values of genotypes and hence adaptation in natural populations (Waddington 1968). Although some of the other participants discussed environmental variability in one context or another, they did not directly consider the problem of how such environments affect the genetic structure of populations.

But perhaps this is not really too surprising in a historical sense. The real impetus to development of this area was, in fact, the magnitude of genetic variation found in natural populations. Up to this time, Levene's model was pretty much restricted to textbooks and other compilations (e.g., Williamson 1958) as yet another example of a mechanism for maintaining polymorphism.

When it was found that an unexpectedly large proportion of the loci in a typical population was segregating for two or more alleles at intermediate frequencies, and that single-locus overdominance explanations were not very satisfactory, the textbooks and compilations were dusted off and the theoreticians began to generate other kinds of models, including a large number dealing with the effects of environmental heterogeneity on genetic structure. We shall not review the development of this theory, since Felsenstein (1976) and Hedrick et al. (1976) have done a nice job. Incidentally, a recent analysis by Maynard Smith and Hoekstra (1980) has led them to conclude that a number of the models are more restrictive than previously thought.

Maynard Smith and Hoekstra notwithstanding, it has become increasingly accepted that environmental variation in both space and time may account for the maintenance of a significant component of the genetic variation in many natural populations (Powell and Taylor 1979). Although a number of correlational studies of genetic and environmental variation have been done (see Nevo 1978), there has been precious little experimental investigation of this theory. This, of course, is a common problem in many areas of population biology (Dawson 1979).

To date, four experimental studies have been reported; all involved measurement of genetic variation at electrophoretically detectable loci and populations exposed to different numbers of variable environmental parameters. In three of the studies a positive relationship between levels of genetic and environmental variability was reported (Powell 1971; McDonald and Ayala 1974; Powell and Wistrand 1978). On the other hand, Minawa and Birley (1978) found no evidence for such a relationship.

The great flurry of interest in explaining existing levels of genetic variation for enzyme loci has tended to obscure the equally paradoxical problem of genetic variability for polygenic traits. It has long been known that natural populations contain large amounts of additive genetic variation for polygenic characters (Lerner 1958; Falconer 1960). Table 6.1 summarizes the quantitative genetic studies that have been carried out with the flour beetles *Tribolium castaneum* and *T. confusum*, the organisms we have used in the experiments to be reported here. It is clear that populations of these beetles are no exception to the general observation noted above. One cannot help but be impressed with the ubiquity of additive genetic variation for a number of components of fitness. Artificial selection studies have commonly produced changes of five standard deviations or more from the means of the foundation populations, and in many of these experiments genetic variation was not exhausted. Clearly there are a large number of loci segregating in both laboratory and natural populations of flour beetles.

Experimental studies with polygenic traits have also demonstrated the widespread occurrence of genotype × environment interactions. These interactions have long been known to be important in populations of domesticated plants and animals. Again, *Tribolium* flour beetles are no exception to this general observation (Sokoloff 1977). For example, artificial selection frequently produces responses that are expressed only in one environment, suggesting that the particular genotypes being selected have extreme values of the selected trait only in certain environments. The existence of extensive genotype–environment interactions is expected if genetic variation is maintained by environmental heterogeneity, since individual genotypes are postulated to have different ranks in phenotypic value in different portions of the range of environments occupied by a population.

Despite an almost astronomical number of artificial selection experiments and other quantitative genetic studies, not much is known about polygenic variation in fitness itself. Genes affecting fitness undoubtedly exhibit considerable pleiotropy. For example, correlated reductions in fitness, which are one possible expression of pleiotropy, are a nearly universal feature of artificial

Table 6.1A. Selection Response and Heritability Estimates of Fitness Components and Body Weight Characters in *Tribolium*.

A. Selection Studies

Species	Character	T	M	DIR	Selection Intensity (%)	h_i^2	GEN	Response (S.D.)	S	Source
CS	Developmental time (egg to pupa)	29	5'	D	.21–.06	.11	13	1.3[a]	—	Dawson 1965a
CS		29	5	U		.32	13	10.9[a]	—	
CF		29	5	D		.15	9	1.3[a]	—	
CS		33	5	D	.18	.38	6	2.0[a]	—	Englert and Bell 1970
CS		33	5	U	.18	.26	6	4.7[a]	—	
CS		32.8	5	D	.40	.26	10	1.9[b]	—	Scheinberg et al. 1967
CS	virgin egg lay	28	5	U	.20	—	20	6.1[b]	F	Orozco 1976
CS		33	5	U	.20	—	20	5.7[b]	F	
CS		38	5	U	.20	—	20	6.4[b]	F	
CS		33	5	U	.05	—	32	7.9[b]	—	Ruano et al. 1975
CS		33	5	U	.10	—	32	7.8[b]	—	
CS		33	5	U	.20	—	32	8.1[b]	—	
CS		33	5	U	.33	—	32	7.4[b]	—	
CS		33	5	U	.50	—	32	5.0[b]	—	
CS		33	5	U	1.00	—	32	0.2[b]	—	
CS		28	5	U	.20	—	9	2.5[b]	F	Orozco and Bell 1974
CS		28	5	U	.20	—	9	3.6[b]	R	
CS		33	5	U	.20	—	9	4.8[b]	F	
CS		33	5	U	.20	—	9	3.9[b]	R	
CS		38	5	U	.20	—	9	4.4[b]	F	
CS		38	5	U	.20	—	9	3.6[b]	R	

Environmental Heterogeneity

	Trait									Reference
CS	14-day larval weight	33	5	U	.37	.16-.28	8	—	—	Wilson et al. 1968
CS		32	5	U	.28	0	14	1.0	—	Okada and Hardin 1970
CS		32	5	D	.28	.17	7	1.1	—	Okada and Hardin 1967
CS	13-day larval weight	32.8	5	U	.40	.30	10	1.9^b	—	Scheinberg et al 1967
CS		33	G^2	U	.28	.31	8	2.8^a	—	Hardin and Bell 1967
CS		33	G	D	.28	.35	8	4.8^a	—	
CS		33	P^3	U	.20	.14	8	2.8^a	—	
CS		33	P	D	.28	.40	8	4.8^a	—	
CS		33	G	U	.20	.30*	16	2.6^d	F	Yamada and Bell 1969
CS		33	G	D	.20	.42*	16	4.2^d	F	
CS		33	P	U	.20	.34*	16	5.7^d	F	
CS		33	P	D	.20	.54*	16	4.1^d	F	
CS		33	GP^4	U	.20	.29*	16	3.1^e	F	
CS		33	GP	D	.20	.37	16	3.4^e	F	
CS		33	G/P^5	U	.20	—	16	3.5^e	F	
CS		33	G/P	D	.20	—	16	3.6^e	F	
CS	19-day pupal weight*	33	5	U	.37	.28-.37*	6	1.8^c	—	Wilson et al. 1965
CS	1st day pupal weight**	32.8	5	U	.40	.30	10	3.4^b	—	Scheinberg et al. 1967
CS		29	5	U	?	.25	22	>7.	—	Wong and Boylan 1970
CS	21-day pupal weight	33	10^5	U	.20	.37**	23	7.5	—	Gall 1971
CS		33	10	D	.20	.26**	25	4.1	—	
CS		33	5	U	.14	—	24	12.0	F	Bell and Moore 1972
CS		33	5	U	.14	—	17	10.0	F	

Table 6.1A. (Cont.)

Species	Character	T	M	DIR	Selection Intensity (%)	h_r^2	GEN	Response	(S.D.)	S	Source
CS	21-day pupal weight	?	?	U	.10	.16	1		.27	—	Meyer and Enfield 1975
CS		?	?	U	.30	.20	1		.19	—	
CS		?	?	U	.50	.14	1		.11	—	
CS		?	?	D	.10	.42	1		.57	—	Meyer and Enfield 1975
CS		?	?	D	.30	.36	1		.36	—	
CS		?	?	D	.50	.21	1		.16	—	
CS		?	?	U	.15	.14	41		—	—	Katz and Enfield 1977
CS		?	?	U	.15	.35***	109		15.0	—	Enfield 1977
CS	30-day adult weight	32	5	U	.28	.16†	14		3.3	—	Okada and Hardin 1970
CS		32	5	D	.28	.35	7		2.4	—	Okada and Hardin 1967

CS = *T. castaneum*; CF = *T. confusum*; DIR = direction of selection (U = up, D = down); GEN = generations of selection for which response is reported; h_r^2 = realized heritability; S = type of selection (I = individual, F = family; R = reciprocal recurrent); T = temperature (°C); M = media

Response (in standard deviations, S.D.)

a. response read from graph; b. response determined as product of average per generation response and total generations of selection; c. average response of 10 lines; d. average of two replicates calculated as in b; e. average of two replicates on G and P media calculated as in b.

Media

1. 5 = wheat flour with the addition of 5% brewer's yeast; 2. G = good media, consisting of corn flour + vitamins + soybean meal + corn oil + brewer's yeast; 3. P = poor media, consisting of corn flour + vitamins + soybean meal; 4 = average weight on G and P was the selection criterion; 5 = weight on G and P measured in alternating generations; 6 = wheat flour with the addition of 10% brewer's yeast.

h_r^2

† mean of two replicates; * range for 10 lines involving mass selection with random mating, assortative mating, disassortative mating, outbreeding, maximum inbreeding; ** h_r^2 for first 10 generations; *** average of two replicates for the first 12 generations; †, average of h_r^2s for first seven and last seven generations.

selection experiments. Because of pleiotropy, gene action with respect to a single character may differ considerably from gene action with respect to fitness. Also, gene effects on the character and fitness scales may be very different.

In artificial selection experiments, the fitness of an individual is determined by the experimenter, and is to a large extent based on phenotypic value for the selected trait. Moreover, since the selection process radically alters the genetic background, observed genotype-environment interactions may not faithfully reflect the true situation in the gene pool of the foundation population.

We have attempted to study the genetic structure of fitness by analyzing a replicated natural selection experiment. Our study was undertaken with the objective of testing the hypothesis that some of the genetic variability underlying continuously distributed traits is maintained by environmental diversity. From assays of polygenic traits in populations that had been allowed to evolve in a series of different environments, we hoped to be able to compare variabilities in populations adapted to simple versus complex environments and to temporally constant versus varying environments.

We shall report an experimental test of an important assumption underlying the models relating genetic variation and environmental heterogeneity. As noted earlier, the models are based on the assumption that different genotypes have highest fitness in the various environmental states occupied by a population, either spatially or temporally. If this is the case, a population established in a single environmental state would be expected to evolve toward the adaptive peak(s) represented by the particular genotype(s) having highest fitness in that environmental state. Moreover, if separate populations were established in different environmental states, they would be expected to diverge genetically during the process of adaptation to the individual environments. If this divergence did not occur, it would be difficult to argue that environmentally diversified selection was responsible for the maintenance of genetic variation in the original populations.

Our experimental material was obtained from a number of populations of the two species, *Tribolium castaneum* and *T. confusum*,

Table 6.1B. Heritability Studies

Species	Character	h^2	Tech	Source
CS	Developmental Time	.26	O/P	Dawson 1965b
CS	(egg-pupa)	.18	HS	
CF		.10	HS	
CS		.15	D	
CF		.11	D	
CS		.11	D	
CS		.43	FS	Scheinberg et al. 1967
CS	Virgin egg lay (28°C)	.27	D/D	Orozco and Bell 1974
CS	(33°C)	.37	D/D	
CS	(38°C)	.32	D/D	
CS	Fecundity: in fresh media	.38	HS	Riddle 1977
CS	in conditioned media	.33	HS	
CF	in fresh media	.34	HS	
CF	in conditioned media	.36	HS	
CS	Egg Number (b population)†	.04	HS	Krause and Bell 1972
CS	(p population)	.01	HS	
CS	Viability (b population)	.07	HS	
CS	(p population)	0	HS	
CS	14-day larval weight	.41	US	Okada and Hardin 1967
CS	13-day larval weight	.59	FS	Scheinberg et al. 1967
CS	13-day larval weight (G)++	.21	HS	Hardin and Bell 1967
CS	13-day larval weight (P)	.23	HS	

h^2 = heritability estimated by one of the following techniques (Tech):
(1) Regression analyses: O/P, offspring-parent; D/D, daughter-dam; D/S daughter-sire; S/D, son-dam; S/S, son-sire; O/S, offspring-sire; IS, intrasire regression of progeny on dam.
(2) Covariance analyses: HS, half sibling; FS, full sibling; US, half sib, with identification of male parent only

that had been maintained without artificial selection in particular environments for over 60 generations. The environmental differences between populations were a series of flour diets utilizing cereal grains that these beetles commonly invade in nature. This environmental variable was chosen because we knew from previous experience that the different flours had large effects on fitness components (see Sokoloff 1974). Also, we knew that the ecology and behavior of these beetles in respect to flour diversity

Table 6.1B. (Cont.)

Species	Character	h^2	Tech	Source
CS	13-day larval weight (G)	.79	FS	
CS	13-day larval weight (P)	.69	FS	
CS	13-day larval weight (G)	.20	IS	
CS	13-day larval weight (P)	.27	IS	
CS	14-day larval weight	.15	S/O	Wilson et al. 1968
CS		.40	D/O	
CS	13-day larval weight (G)	.37	HS	Yamada and Bell 1969
CS	(P)	.43	HS	
CS	(b population)	.18	HS	Krause and Bell 1972
CS	(p population)	.05	HS	
CS	larval number (b population)	.33	D/D	
CS	(p population)	.12	D/D	
CS	21-day pupal weight	.46	O/S	Bell and Moore 1958
CS		.33	HS	Boylan and Wong 1965
CS		.34	S/S	Enfield et al. 1966
CS		.21	HS	Katz and Enfield 1977
CS		.24	HS	Kaufman et al. 1977
CS		.27	O/P*	
CS	Emigration rate	.71	HS**	Riddle 1977
CF	Emigration rate	.86	HS**	

(3) Diallel analysis, D
*Pooled value for S/S, S/D, D/S and D/D regressions
**Based on full sib family means.
+ b and p were two unrelated laboratory populations
++ G = good media, P = poor media (see part A of table)

in nature were favorable to the maintenance of genetic variation. If diversifying selection in natural populations is responsible for maintaining genetic variation for utilization of different types of flours, we reasoned that we should be able to observe adaptation to given flour combinations in laboratory populations.

The division of responsibility for the work we are reporting is this: the populations were initiated and maintained by PSD, the analytical experiments and analyses were carried out by RAR, and primary responsibility for the manuscript and its delivery at the Colloquium was undertaken by PSD. Our results are presented in more detail in Riddle, Dawson and Zirkle (1983).

The Populations and the Experiments

The experimental populations were established in 1969. A total of 32 populations, including two replicates of each species for each of eight selection treatments, were set up as shown in table 6.2. Each experimental population was initiated with a random sample of 500 adults from newly synthesized populations of the two species. The latter populations, designated as the Oregon synthetic populations, were in turn constructed using samples of beetles obtained by PSD from a number of different laboratories. Some were long-time laboratory populations and others had been in the laboratory only a short time since being captured from nature. The rationale for building these new synthetic populations was to have highly heterogeneous gene pools in the foundation stocks in order to maximize the probability of being able to detect adaptation to the experimental treatments. One series, or selection treatment, was started each week over an eight-week period.

The populations were transferred every eight weeks, at which

Table 6.2. The Experimental Populations

Flour		Tribolium confusum		Tribolium castaneum	
Type*	Series	Rep. 1	Rep. 2	Rep. 1	Rep. 2
C	1	1-1	1-2	1-3	1-4
W	2	2-1	2-2	2-3	2-4
CW	3	3-1	3-2	3-3	3-4
WC	4	4-1	4-2	4-3	4-4
WCBR	5	5-1	5-2	5-3	5-4
R	6	6-1	6-2	6-3	6-4
B	7	7-1	7-2	7-3	7-4
W/C	8	8-1	8-2	8-3	8-4

* C = corn, W = wheat, CW = 2/3/ corn + 1/3 wheat, WC = 2/3 wheat + 1/3 corn, WCBR = 1/4 wheat + 1/4 corn + 1/4 barley, + 1/4 rye, R = rye, B = barley, W/C = alternating wheat and corn.

time the adult beetles were discarded and the pupae, larvae, and eggs were placed in a fresh medium. This procedure was adopted in order to insure that there would be at least one generation every eight weeks. For series 8, the type of flour was changed at each transfer.

During the 10 plus years of their existence the populations have been in four different incubators at two universities. However, all 32 populations have been kept together at all times and have therefore been exposed to the same environments.

The analytical experiments to be discussed in this paper were carried out using animals derived from individuals taken from the experimental populations at transfers 58-60. Details of the procedures utilized may be found in Riddle, Dawson, and Zirkle (1983). We established a total of 224 test populations, representing each of the 32 original populations grown for two generations on each of the seven flour diets. By growing the test populations for two generations on the diets to be used for our experiments, we eliminated environmentally induced maternal effects as a source of variation.

Our experimental design initially involved measuring various fitness components in each of the 224 test populations on each of the four single flours (corn, wheat, rye, barley). After looking at the results from the first set of experiments with *T. confusum*, we revised our plans. All subsequent experiments with *T. confusum* and all experiments with *T. castaneum* were done using only the populations from series 1, 2, 3, 4, and 8 grown on C, W, CW, and WC flour types (see table 6.2 for explanation of symbols).

We studied a variety of growth and fitness characters. Survival was measured in *T. confusum* only, from the egg stage to several larval ages. Larval age was not a significant source of mortality, so the data were pooled to provide a single measure of survival. A second measure of survival, from the egg to the pupal stage, was obtained in a later experiment.

All remaining characters were measured in both species. Larval growth rate was calculated as the regression coefficient from regression of the logarithm of larval weight on age. Absolute larval weight, which was not highly correlated with larval growth rate, was analyzed separately. Pupal weight, which is not highly correlated with larval weight but is very highly correlated with adult weight (Englert and Bell 1963, 1969), was obtained at a constant age following pupation. Developmental time was measured from day of egg laying until pupation on cohorts obtained from a 24-hour egg collection. Fecundity was measured at two densities: 1 beetle/gram and 3.33 beetles/gram, referred to as low and high density respectively. In the low-density experiments, net fecundity was measured on single females placed in 1 gram of medium for a 48-hour period. In the high-density experiments, groups of 10 males and 10 females were placed in 6 grams of medium for 24 hours. For the latter experiments, the test vials contained 50 eggs, marked with neutral red dye and randomly distributed in the vials according to the technique invented by Rich (1956). This allowed direct measurements of egg cannibalism rates as well as estimates of real fecundity. The high-density experiment was also repeated using conditioned wheat and corn flours, providing estimates of real fecundity and cannibalism rates in conditioned media.

The data were analyzed by factorial analysis of variance involving series and flour type (selection treatment). Evidence for adaptation to the specific flour types in which selection had occurred would be expressed as a significant interaction between the two main effects. Replicates were treated as a random factor nested within series; this yielded a mixed model for analysis.

The Data

Before briefly presenting the results of comparative analyses of the experimental populations, we shall give evidence that the flour types affect our beetles and that the populations have differentiated in response to flour type. Figures 6.1 and 6.2 contain data on three-week larval weights in our outbred Oregon populations on each diet and for the series populations on the selection diets. These data are based on 80 to 100 larvae weighed in groups of twenty.

Environmental Heterogeneity

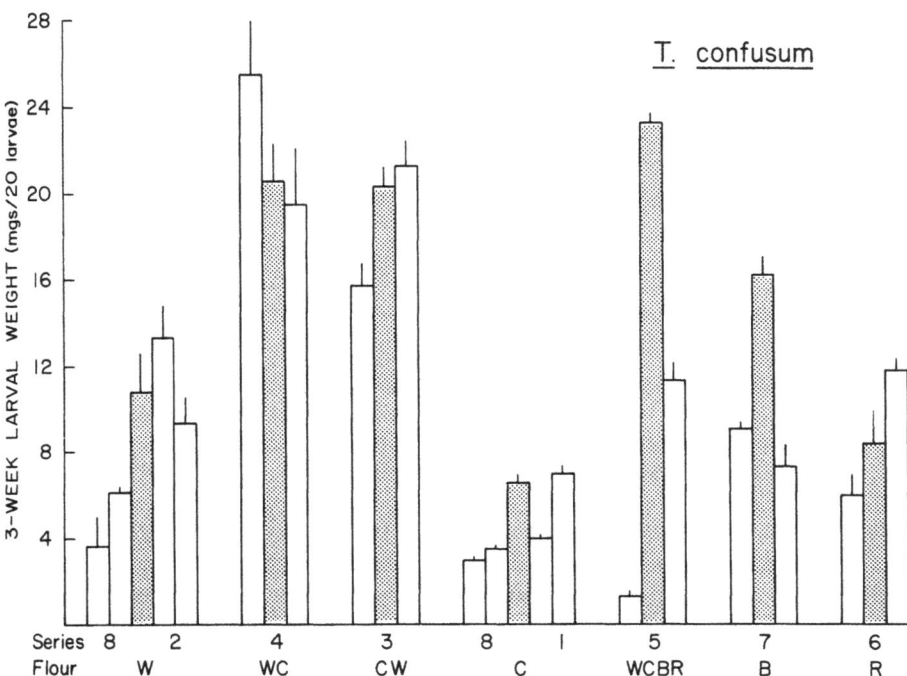

Figure 6.1. Three-week larval weights in *T. confusum* from the Oregon population (dotted bars) and the series populations reared on the designated flour types.

(See table 6.2 for abbreviations). Each series has two replicates. Vertical lines are standard errors.

First, flour has a significant effect on weight in the Oregon populations of both *T. castaneum* and *T. confusum*. This clearly shows that the diets have different biochemical and physiological effects on flour beetles. In addition, the effect of flour type is species specific. For *T. confusum* (figure 6.1), growth is significantly greater on the mixed diets than on single flour types. For *T. castaneum* the mixed diets do not increase growth rates; however, growth decreases with increasing amounts of corn flour in the diet.

Comparing the experimental (series) populations with the Oregon populations shows that they tend to be smaller on their selection diets. The most extreme example of this is population 5-1,

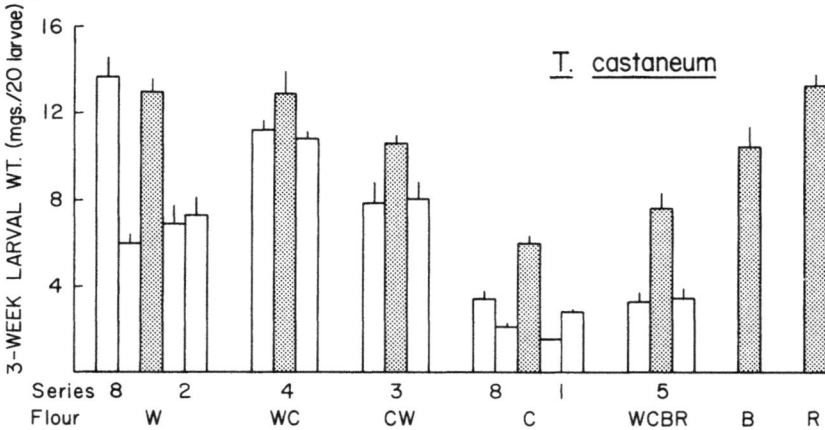

Figure 6.2. Three-week larval weights in *T. castaneum*. Same as figure 6.1.

in which three-week larval weight is only 1/20 of that in the Oregon population of *T. confusum* (figure 6.1). Another point of interest is that replicate populations are different in many cases.

Although it is clear that the populations have diverged genetically, the changes were not those that we originally expected. Three-week larval weight reflects some combination of consumption rates and digestive efficiencies, and we had expected natural selection to act on one or both of these characters, leading to higher growth rates. That the populations have lower three-week weights on the flour types they have experienced through time than does our outbred population, which has never been exposed to barley, rye, or corn, suggests that all other things are not equal. That is to say, the observed genetic changes were in the opposite direction to those we expected with respect to this trait. Other selective forces must have been in operation in these populations.

All subsequent experiments involved all populations or a subset of the populations reared on several flour types. As explained earlier, we were looking at the flour × series interaction from a factorial analysis of variance. Flour × series interactions are expected to be nonsignificant on the basis of the null hypothesis of no adaptation to the specific selection diets. This provides our test of the hypothesis that populations maintained in single environmental states are expected to diverge as a result of natural selection for adaptation to those environments.

A very brief summary of the analyses of data obtained from these experiments is presented in table 6.3. Blocks, sexes, and other sources of variation incidental to our interests in the present context are not shown. The results can be summarized as follows:

(1) The flour × series interaction was not a significant source of variation in any of the 20 analyses.
(2) Flour exhibited highly significant effects ($P < 0.001$) on many traits. This provides further evidence for differences among flour types on the physiology and reproduction of the beetles.
(3) Replicate differences were highly significant for 15 out of 20 characters, whereas series effects were significant in only three instances. Replicate differences are evidence of genetic differentiation although not of a type related to the environmental treatments.

Table 6.3. Summary of Analyses of Variance of Components of Growth and Fitness in *Tribolium confusum* and *T. castaneum*.

	Significance Levels of Mean Squares										
Source	LS_1	LS_2	GR	LW	DT	PW	FL	FH	FC	CF	CC
T. confusum											
Series (S)	NS	***	NS	NS	NS	NS	*	NS	NS	NS	NS
Replicates (R)/S	***	NS	***	***	***	***	***	***	NS	***	NS
Flour (F)	***	***	***	***	NS	NS	NS	NS	NS	NS	NS
F × S	NS	NS	NS	NS	NS	NS	NS	NS	NS	NS	NS
F × R/S	***	NS	—	**	***	***	***	NS	NS	***	NS
T. castaneum											
Series (S)	———	**	NS	NS	NS	NS	NS	NS	NS	NS	NS
Replicates (R)/S	———		NS	***	**	***	***	***	NS	***	***
Flour (F)	———		***	NS	***	NS	NS	NS	*	***	***
F × S	———		NS	NS	NS	NS	NS	NS	NS	NS	NS
F × R/S	———		—	NS	***	***	***	*	**	NS	NS

NS, $P > 0.05$; *, $P < 0.05$; **, $P < 0.01$; ***, $P < 0.001$
NOTE: Abbreviations are as follows: LS_1 and LS_2, larval survival; GR, growth rate; LW, three week larval weight; DT, developmental time; PW, pupal weight; FL, fecundity at low density (single females); FH, fecundity at high density; FC, fecundity at high density in conditioned flour; CF, cannibalism rate in fresh flour; CC, cannibalism rate in conditioned flour. Nested factors are indicated in the first column by a "/".

(4) Flour × replicate interactions were significant in 11 of 18 analyses. These interactions demonstrate that sets of populations do not necessarily respond similarly to the flour differences and therefore are evidence of genotype × environment interactions.

Discussion

The hypothesis that a significant component of the genetic variation in natural populations is maintained by diversifying selection in a heterogeneous environment is not supported by our results. We had expected to find that populations isolated on various flour types would adapt to the different media, and that this adaptation would be expressed by "better" growth and reproduction by a given population on the flour type on which it had evolved through time. To us "better" would presumably mean faster, or more, such as faster growth, faster development, larger size, higher fecundity, or higher survival. Since we failed to detect significant differentiation among our populations on their selected flour types, we are forced to conclude that our foundation population did not contain a significant level of additive genetic variation that could be utilized to adapt to these naturally occurring flour types.

This lack of genetic variation in fitness is surprising, since *Tribolium* are known to live in the various flours in nature. Moreover, King and Dawson (1973) and others have shown that both adults and larvae are capable of habitat selection. As Powell and Taylor (1979) have recently argued, habitat selection behavior should enhance the maintenance of genetic variation by environmentally diversified selection. The differences between flour types in our experiment were apparently sufficient to prevent adaptation by natural selection.

This of course is not to say that additive genetic variation for *individual* fitness components was not present in the foundation populations. First, we have good evidence of divergence in the experimental populations (figures 6.1 and 6.2). Second, we have indirect evidence from measures of additive genetic variation for female pupal weight in several of the *T. confusum* populations (Zirkle and Riddle, 1982). Third, table 6.3 demonstrates that flour × replicate interactions were significant in most of the analyses. Clearly there were extensive genotype × environment interactions. In fact these interactions were generally of the same magnitude as the flour × series interactions. In our experimental design replicates were initially established to serve as a measure of random divergence between populations subjected to the same environmental treatment (flour type). Thus replicates were treated statistically as a random factor, and differences between replicates are ascribed to random genetic drift. However, it is possible to view the replicates in another way. In an adaptive landscape with multiple peaks representing multi-locus genotypes with high fitness, replicate populations may follow different trajectories in adapting to the environmental treatments. In this case, replicate differences may not simply reflect random noise, but would result from the combined effects of both selection and random genetic drift. Replicates would then be a fixed effect, and both the flour × series and flour × replicate interactions would be tested over the same error mean square. Flour × series interactions would then also be significant in many of the analyses. At any rate, this reinforces the notion that genotype × environment interactions are widespread, and increases our initial surprise in finding no evidence for response to selection in spite of the presence of variation in fitness components.

Why have we failed to detect adaptation to the various flour types even though we have evidence for additive genetic variation in fitness characters? One possible explanation is that we did not analyze the correct traits. But this explanation is countered by evidence from our studies. For example, the flour types do have significant effects on the beetles, as shown in figures 6.1 and 6.2 for larval weights. This was true for many of the traits we studied, therefore it is difficult to argue that we did not provide different environments or that the different environments did not affect the beetles. In addition, as discussed above, there is good evidence for genetic changes in the traits we did measure. And finally, we know from other investigations with *Tribolium* that artificial selection in different nutritional environments has been successful (e.g., Hardin and Bell 1967; Yamada and Bell 1969). We are confident that we provided the opportunity for adaptation to occur and that we measured the appropriate characters to detect such adaptation.

A second explanation for lack of adaptation is based on random effects associated with population size. However, we were unable to demonstrate any relationship between population means for the characters we measured and population size (Riddle, Dawson, and Zirkle 1983).

We are left with the firm conviction that "something" about the genetic structure of fitness in these populations has prevented them from responding to selection. We propose that the "something" is pleiotropic effects of genes on different fitness characters which are expressed as negative correlations among traits. Pleiotropic gene effects are an almost universal feature of gene expression, as forcefully pointed out by Ernst Caspari some 30 years ago (Caspari 1952). Since that time relatively little attention has been paid to pleiotropy, although it is an integral part of Sewall Wright's shifting balance theory of evolution (Wright 1977, 1978).

When we compared the replicate populations from the same flour type, there were sometimes reversals of ranks for different fitness components (Riddle, Dawson, and Zirkle 1983). These could be explained by pleiotropy or by inbreeding depression, resulting from fixation of different deleterious alleles in the two replicates. Since, as noted earlier, we did not find any relationship between the population means for these traits and population size, we favor an explanation based on pleiotropy.

In our experimental populations, natural selection was based on overall fitness. Thus, the frequent observation from artificial selection experiments that pleiotropic effects of genes slow down progress only in later generations is not at odds with our results. This is because in such experiments individuals are "selected" on the basis of a single character. The negative effects of alleles on other traits may not become an obvious problem until after some generations of selection.

Why do we believe that the environmental treatments (flour types) were sufficiently different to prevent adaptation by natural selection? Or, to say it another way, why do we think that diversifying selection for adaptation to these environmental differences is not responsible for a significant component of the genetic variation present in natural populations? Our response is that the differences between flour types are actually quite complex in comparison to the biochemical and behavioral plasticity of the beetles. In Dick Levins' (1968b) terms, the environmental differences are large relative to the tolerance of an individual phenotype. In contrast to previous studies involving nutritional environments, the flours we used were not supplemented with brewer's yeast. This was done to make the environments more realistic. However, it is known that yeast supplementation has manifold nutritional effects, which would obscure many differences between flours. There are many differences in composition of our four flours for amino acids, vitamins, and minerals. For example, the balance of amino acids is quite different (Riddle, Dawson, and Zirkle 1983). Thus we believe that the complexity of differences among the nutritional regimes used to maintain the experimental populations for over 10 years were sufficiently different to prevent differentiation of our populations. That is, negative correlations among the fitness components resulting from pleiotropic gene effects were sufficiently important that changes in individual components were opposed by correlated changes in others. Thus we cannot account for the maintenance of genetic variation for utilization of these flour types in natural populations by environmentally diversified selection.

Pleiotropic interactions, expressed as fitness reversals between characters, provide a mechanism to explain the apparently contradictory results of our experiment. When genotypic contributions to fitness are averaged across all characters, fitness differences may disappear or even exhibit overdominance. An example of this effect has been reported by Marinkovic and Ayala (1975). If pleiotropic effects interact in this way, the apparently conflicting observations that genotype × environment interactions were common for all components and that adaptation did not occur are compatible. Furthermore, pleiotropic interactions in a multilocus system provide a mechanism to generate multiple-peak epistasis, so that differentiation of replicates can be explained as an interaction between natural selection and random genetic drift.

The results of our experiment suggest that the theoretical basis for models of environmentally diversified selection may be inadequate. Pleiotropic effects are not explicitly considered in these models nor is the relationship among various environmental variables and biological variables considered. Our data suggest that all components of "environmental heterogeneity" do not have equivalent effects on genetic parameters of a population, such as levels of variability. These results present a challenge to theoreticians to incorporate the biologically real phenomenon of pleiotropy and the ecologically real phenomenon of complex and conflicting selective pressures into the theoretical structure of population biology.

ACKNOWLEDGMENTS

This work was supported by the National Science Foundation (DEB-7725569). We thank Dave Zirkle for helpful comments on the manuscript.

REFERENCES

Bell, A. E. and C. H. Moore. 1958. Further comparisons of reciprocal recurrent selection with conventional methods of selection for improvement of quantitative characteristics. *Proc. X Intern. Congr. Genet.* 2:20-21.

Bell, A. E. and C. H. Moore. 1972. Reciprocal recurrent selection for pupal weight in *Tribolium* in comparison with conventional methods. *Egypt J. Genet. Cytol.* 1:92-119.

Boylan, W. J. and W. C. Wong. 1965. Responses to selection and combining ability in *Tribolium castaneum*. *Genetics* 51:429-430.

Caspari, E. G. 1952. Pleiotropic gene action. *Evolution* 6:1-18.

Dawson, P. S. 1965a. Genetic homeostasis and developmental rate in *Tribolium*. *Genetics* 51:873-885.
Dawson, P. S. 1965b. Estimation of genetic components of phenotypic variance for developmental rate in *Tribolium*. *Heredity* 20:403-417.
Dawson, P. S. 1979. Evolutionary changes in egg-eating behavior of flour beetles in mixed species populations. *Evolution* 33:585-594.
Dempster, E. R. 1955. Maintenance of genetic heterogeneity. *Cold Spring Harbor Symp. Quant. Biol.* 20:25-32.
Enfield, F. D. 1977. Selection experiments in *Tribolium* designed to look at gene action issues. In E. Pollak, O. Kempthorne, and T. B. Bailey, Jr., *Proceedings of the International Conference on Quant. Genet.* Ames: Iowa State University Press.
Enfield, F. D., R. E. Comstock, and O. Braskerud. 1966. Selection for pupa weight in *Tribolium castaneum*. I. Parameters in base populations. *Genetics* 54:523-533.
Englert, D. C. and A. E. Bell. 1963. Genetic differences in the growth curve of *Tribolium castaneum*. *Growth* 27:87-99.
Englert, D. C. and A. E. Bell. 1969. Components of growth in genetically diverse populations of *Tribolium castaneum*. *Can. J. Genet. Cytol.* 11:896-907.
Englert, D. C. and A. E. Bell. 1970. Selection for time of pupation in *Tribolium castaneum*. *Genetics* 64:541-552.
Falconer, D. S. 1960. *Introduction to Quantitative Genetics*. New York: Ronald Press.
Felsenstein, J. 1976. The theoretical population genetics of variable selection and migration. *Ann. Rev. Genet.* 10:253-280.
Gall, G. A. E. 1971. Replicated selection for 21-day pupa weight of *Tribolium castaneum*. *Theor. Appl. Genet.* 41:164-173.
Hardin, R. T. and A. E. Bell. 1967. Two way selection for body weight in *Tribolium* on two levels of nutrition. *Genet. Res., Camb.* 9:309-330.
Harris, H. 1966. Enzyme polymorphisms in man. *Proc. Roy. Soc. B* 164:298-310.
Hedrick, P. W., M. E. Ginevan, and E. P. Ewing. 1976. Genetic polymorphism in heterogeneous environments. *Am. Rev. Ecol. Syst.* 7:1-32.
Katz, A. J. and F. D. Enfield. 1977. Response to selection for increased pupa weight in *Tribolium castaneum* as related to population structure. *Genet. Res., Camb.* 30:237-246.
Kaufman, P. K., F. D. Enfield and R. E. Comstock. 1977. Stabilizing selection for pupa weight in *Tribolium castaneum*. *Genetics* 87:327-341.

King, C. E. and P. S. Dawson. 1973. Habitat selection by flour beetles in complex environments. *Physiol. Zool.* 46:297-309.
Krause, E. and A. E. Bell. 1972. A genetic study of biomass in *Tribolium*. I. Path coefficient analysis of base populations. *Can. J. Genet. Cytol.* 14:181-193.
Lerner, I. M. 1958. *The Genetic Basis of Selection.* New York: John Wiley and Sons.
Levene, H. 1953. Genetic equilibrium when more than one ecological niche is available. *Am. Nat.* 87:331-333.
Levins, R. 1968a. Evolutionary consequence of flexibility. In R. C. Lewontin, ed. *Population Biology and Evolution.* Syracuse: Syracuse University Press.
Levins, R. 1968b. *Evolution in Changing Environments.* Princeton: Princeton University Press.
Lewontin, R. C., ed. 1968. *Population Biology and Evolution.* Syracuse: Syracuse University Press.
Lewontin, R. C. and J. L. Hubby. 1966. A molecular approach to the study of genic heterozygosity in natural populations. II. Amount of variation and degree of heterozygosity in natural populations of *Drosophila pseudoobscura. Genetics* 54:595-609.
Marinkovic, D. and F. J. Ayala. 1975. Fitness of allozyme variants in *Drosophila pseudoobscura.* I. Selection at the PGM-1 and ME-2 loci. *Genetics* 79:85-95.
Maynard Smith, J. and R. Hoekstra. 1980. Polymorphism in a varied environment: how robust are the models? *Genet. Res.* 35:45-57.
McDonald, J. F. and F. J. Ayala. 1974. Genetic response to environmental heterogeneity. *Nature* 250:572-574.
Meyer, H. H. and F. D. Enfield. 1975. Experimental evidence on limitations of the heritability parameter. *Theor. Appl. Genet.* 45:268-273.
Minawa, A. and A. J. Birley. 1978. The genetical response to natural selection by varied environments. *Heredity* 40:39-50.
Nevo, E. 1978. Genetic variation in natural populations: patterns and theory. *Theoret. Popul. Biol.* 13:121-177.
Okada, I. and R. T. Hardin. 1967. An experimental examination of restricted selection index, using *Tribolium castaneum.* I. The results of two-way selection. *Genetics* 57:227-236.
Okada, I. and R. T. Hardin. 1970. An experimental examination of restricted selection index using *Tribolium castaneum.* II. The results of long-term one-way selection. *Genetics* 64:533-539.
Orozco, F. 1976. A dynamic study of genotype-environment interaction with egg laying of *Tribolium castaneum. Heredity* 37:157-177.

Orozco, F. and A. E. Bell. 1974. A genetic study of egg laying of *Tribolium* in optimal and stress environments. *Can. J. Genet. Cytol.* 16:49–60.
Powell, J. R. 1971. Genetic polymorphisms in varied environments. *Science* 174:1035–1036.
Powell, J. R. and C. E. Taylor. 1979. Genetic variation in ecologically diverse environments. *Am. Sci.* 67:590–596.
Powell, J. R. and H. Wistrand. 1978. The effect of heterogeneous environments and a competitor on genetic variation in *Drosophila*. *Am. Nat.* 112:935–947.
Rich, E. R. 1956. Egg cannibalism and fecundity in *Tribolium*. *Ecology* 37:109–120.
Riddle, R. A. 1977. Density dependent selection and the maintenance of genetic variation in life-history attributes: Evidence from *Tribolium castaneum* and *T. confusum* in temporally heterogeneous environments. Ph.D. thesis. Oregon State University.
Riddle, R. A., P. S. Dawson and D. F. Zirkle. 1983. An experimental test of the environmental variation-genetic variation relationship using *Tribolium*. *Genetics* (in press).
Ruano, R. G., F. Orozco and C. López-Fanjul. 1975. The effect of different selection intensities on selection response in egg-laying of *Tribolium castaneum*. *Genet. Res., Camb.* 25:17–27.
Scheinberg, E., A. E. Bell and V. L. Anderson. 1967. Genetic gain in populations of *Tribolium castaneum* under uni-stage tandem selection and under restricted selection indices. *Genetics* 55:69–90.
Sokoloff, A. 1974. *The Biology of Tribolium*. Vol. 2. Clarendon Press, Oxford.
Sokoloff, A. 1977. *The Biology of Tribolium*. Vol. 3. Clarendon Press, Oxford.
Waddington, C. H. 1968. The paradigm for the evolutionary process. In R. C. Lewontin, ed. *Population Biology and Evolution*. Syracuse: Syracuse Univ. Press.
Williamson, M. H. 1958. Selection, controlling factors and polymorphism. *Am. Nat.* 92:329–335.
Wilson, S. P., W. H. Kyle and A. E. Bell. 1965. The effects of mating systems and selection on pupa weight in *Tribolium*. *Genet. Res., Camb.* 6:341–352.
Wilson, S. P., P. V. Blair, W. H. Kyle and A. E. Bell. 1968. The influence of selection and mating systems on larval weight in *Tribolium*. *J. Heredity* 59:313–317.
Wong, W. C. and W. J. Boylan. 1970. Intrapopulation selection and cor-

related response in crossbreds of *Tribolium castaneum*. *Genetics* 64: 69-78.

Wright, S. 1977. *Evolution and the Genetics of Populations*, Volume 3. Univ. of Chicago Press.

Wright, S. 1978. *Evolution and the Genetics of Populations*, Volume 4. Univ. of Chicago Press.

Yamada, Y. and A. E. Bell. 1969. Selection for larval growth in *Tribolium* under two levels of nutrition. *Genet. Res., Camb.* 13:175-195.

Zirkle, D. F. and R. A. Riddle. 1982. Quantitative genetic response to environmental heterogeneity. *Evolution* (in press).

7 Plant Parentage: An Alternate View of the Breeding Structure of Populations

DONALD A. LEVIN

The breeding structure of populations is of central importance in evolutionary genetics since it dictates, in part, the relatedness of gametes being joined during fertilization, the genetic structure of zygote populations, the spatial organization of variation, and levels of gene pool homogenization. Plant evolutionists have sought to characterize population breeding structure from pollen and seed dispersal patterns, and from family genotype distributions. Dispersal patterns are used for estimating neighborhood size (Levin and Kerster 1974), and genotype distribution for estimating the level of random outcrossing (Brown et al. 1975; Clegg et al. 1978). These estimates speak to the genetic structure of the seed phase of the life cycle. If one wishes to interpret the genetic structure of adult populations, one could use the neighborhood or outcrossing estimates. However, genotype-dependent seed germination and establishment, and plant survivorship, may significantly alter the structure of populations. Consider for example a population of annual self-compatible plants. If the products of selfing had a significantly lower survivorship than products of outcrossing, the adult population would be much more outbred than the seed population from which it was derived.

The genetic structure of mature plants, expressed in terms of inbreeding parameters, may be obtained from a understanding of plant parentage. Reproductive plants have egg and pollen parents. Parents have a relationship in space which is influenced by pollen dispersal patterns, cross-compatibility relationships, and seed and plant survivorship. If we knew the identity of parents and their genotypes, we would be able to make a definitive statement about the realized breeding structure of populations. Indeed it would be more informative than estimates from gene flow or progeny analyses, since we would be dealing with plants actually established rather than seeds still associated with female parents. In the absence of parental identity, the distribution of parents in space offers insight into the realized breeding structure, much as the distribution of mating plants offers insight into the mating pattern.

Pollen Flow vs. Pollinator Movements

We might expect interparent distances to be distributed as pollinator flight distances. This would lead to the conclusion that egg and pollen parents often are neighboring plants (Levin and Kerster 1974). The tacit assumption in employing pollinator data in this way, is that all pollen from one plant is deposited on the next plant visited. If it is not—i.e., if there is pollen carryover—the pollen dispersal distances will exceed interplant pollinator flight distances. The magnitude of pollen carryover has become evident only recently. We established a core (several flats) of a cultivar homozygous for a dominant marker in a population of *Phlox drummondii*. Seeds were collected in concentric rings at distances from the core, grown in the greenhouse, and hybrids were identified. Pollinator flight distances were observed at the study site. We would expect a one-to-one correspondence between flight distance and gene dispersal if pollen were deposited on the first plant visited. Mean dispersal distance for genes was 1.5 times that for pollinator flights (Levin 1981). The sole explanation for this disparity is pollen carryover. Schaal (1980) reported a disparity of similar magnitude between gene dispersal and bee flight distance in a synthetic population of *Lupinus texensis*. Again, pollen carryover is the only logical explanation. Thomson and Plowright (1980) studied pollen carryover in bee-pollinated plants by counting the numbers of pollen grains applied to stigmas of a series of flowers. In general, most grains were deposited on the first few flowers, but some were carried to the twentieth flower or beyond. Of additional interest was their observation that pollen deposition per flower is a positive function of the amount of nectar present. Pollen carryover is increased when flowers offer little reward; pollinators spent little time on them. In summary, interparent distances may be considerably greater than pollinator foraging patterns suggest.

Differential Crossability

It is tempting to assume that parental distances are distributed as are the distances of pollen grains. However, pollen grains from all plants do not have an equal probability of effecting fertilization in a given plant. The self-incompatibility (S) locus may preclude certain types of crosses. Pollen which shares a self-incompatibility allele with another plant is excluded from breeding with it (de Nettancourt 1977). Given the richness of S-alleles in natural populations, plants sharing S-alleles are likely to be related. This means that in populations where seed dispersal is highly restricted, and thus where neighboring plants are often related, a distant pollen grain is more likely to effect fertilization than a pollen grain from a neighboring plant.

Given that pollen-pistil compatibility might vary with distance, we sought such a relationship in populations of *Phlox drummondii*. Pollen and seed dispersal in this species typically is a few meters or less, and populations display significant heterozygote deficiency relative to Hardy-Weinberg expectations (Levin 1977). These observations suggest that populations undergo moderate inbreeding, and raise the possibility of proximity-dependent crossing effectiveness. This possibility was addressed by analyzing (1) the percentage of pollen grain germination on stigmas, (2) percentage of seed abortion and (3) percentage of ovules developing into sound seed as functions of distance within and between populations. In a pilot study, plants were collected as seedlings along 35 m linear transects in each of 14 populations and were grown to maturity in the greenhouse. Crosses were made within and between populations. Crossability was considered in terms of the spatial relationships of the pollen and egg parents, which were known. On the average, pollen grain germination increased with distance up to 10 m. Additional distances up to 35 m. were not accompanied by overall changes in germination percentage. Pollen germination averaged about 17 percent with near-neighbor pollinations vs. ca. 20 percent with pollinations involving plants 10 m. away or more. Pollen from parents in neighboring populations displayed germination rates similar to those found in intrapopulation crosses beyond 10 m. Pollen from sources greater than 20 km away performed substantially below that of pollen from within populations or from neighboring populations, mean germination being ca. 19 percent and 14 percent, respectively. Thus pollen germination rates are highest when pollen parents are moderate distances away from egg parents.

The quality of pollen parents, expressed in terms of pollen quantity and fertilization ability clearly varies among plants as a function of their genotypes. In maize, pollen quantity per plant varies between inbred lines, and the total number of pollen grains in hybrids may greatly exceed that of inbred lines (Pfahler 1970). The vigor of pollen—i.e., it germinability and growth rate—also varies in maize between inbred lines (Pfahler 1965; Gorla et al. 1975; Johnson and Mulcahy 1978). Moreover, Murakami et al. (1972) found that F_1 hybrid pollen frequently grows faster than pollen from inbred parents. Pollen grains from the same genetic lineage as the egg parent may be less effective in fertilizing than pollen grains from genetic lineages unrelated to the egg parent (Pfahler 1965). Negative assortative mating is thereby promoted in mixed populations. The extent to which crosses between families, or between hybrids and inbred plants, are favored relative to crosses among relatives remains to be determined in natural populations. However, the pattern in maize is what we might expect to find in nature, and the consequence is a more open breeding structure.

The quality of pollen parents vis à vis a given single-egg parent depends not only on cross-compatibility, but also upon the vigor of the progeny before and after dispersal. Pre-dispersal weakness, which is genotype-dependent, may be manifested as seed abortion; the developing seed is sensitive to genetic disharmonies brought about by inbreeding and wide hybridization. High levels of seed abortion following inbreeding are well known in several predominantly outcrossing crop species and in conifers (Crumpacker, 1967; Franklin 1970). Native angiosperms have received less attention in this respect.

We can investigate the consequences of severe inbreeding in self-incompatible species through a "mentor-pollen technique" as demonstrated by Dayton (1974) and Pandey (1977). In this method, compatible (outcross) pollen which has been "killed" by repeated freezing and thawing, chemical treatment, or irradiation is mixed with live self-pollen. A pollen tube growth promoting substance, or a regulator for the production of this substance, is released from the "killed" pollen and stimulates the growth of self pollen (Pandey 1979). This mentor technique was used to overcome self-incompatibility in *Phlox drummondii* collections obtained throughout the range of the species. The incidence of abortion in selfed seed varies from 20 to 45 percent between populations, roughly half being a result of the seed genotype.

Of particular interest within the context of this discussion is whether abortion is dependent on the distance between egg and pollen parents. If relatedness is a function of distance, we would expect abortion levels between neighboring plants to be higher than that between distant plants. In populations of *Phlox drummondii*, seed abortion decreased with interparent distances up to about 10 m, and remained at that level as distances increased to 35 m. Abortion following crosses of 1 m or less averaged 18 percent vs. 14 percent following crosses of 10 m+. It seems likely that higher abortion in the progeny of crosses between neighbors reflects their overall relatedness, because abortion levels increase almost 30 percent with self-fertilization. Crosses between neighboring populations did not yield lower abortion levels than that observed at interparent distances beyond 10 m within populations. Crosses between populations more than 20 km apart did yield significantly higher abortion rates (19 percent vs 14 percent). Distance-dependent abortion rates also have been described within *Picea glauca* (Coles and Fowler 1976). Crosses between plants less than 100 m apart yielded 28 percent less sound seed than crosses at greater distances. Selfing produces little sound seed. Presumably some plants near each other were relatives, so that inbreeding varied as a function of distance.

Seed-set in plants is dependent upon pollen-pistil compatibility and seed variability. In *Phlox drummondii*, seed-set tends to increase as the distance between plants within a population increases. This pattern is evident across populations. Seed-set from crosses between near neighbors averages 53 percent compared with 65 percent for crosses between plants at least 10 m apart. Crosses between neighboring populations have seed-sets similar to that of crosses between plants 20 m apart ($\bar{x}$ = 63%). However, crosses between populations more than 20 km apart yielded fewer seeds per ovule ($\bar{x}$ = 55%) than crosses between neighboring populations. Thus the general pattern is one of an intermediate optimum, more than a few meters and less than 20 km.

Seed-set following crosses between parents separated by various distances has also been analyzed in *Delphinium nelsoni* by Price and Wasser (1979). Interparent distances analyzed were 1 m, 3 m, 30 m, 100 m, and 1000 m. For the most part, 10 m crosses gave the highest seed-set in the two study populations over two years. The pattern of an intermediate optimum also has been described in *Stylidium*. In *S. elongatum* and *S. confluens*, the percentage seed-set was greater in crosses separated by 40 to 60 km than in crosses of smaller or greater distances (Banyard and James 1979). Since embryo abortion often follows self-fertilization in these species, it seems likely that reduced seed-set in intrapopulation crosses was a manifestation of inbreeding depression, as was the case in *Phlox*. A decline in seed-set between very distant populations may be due to a loss of coadaptation fostered by divergent local environmental adaptations.

There are several factors in addition to proximity which determine the likelihood of a pollen parent crossing with a given egg parent. The most obvious factor is time of flowering. The greater the overlap in the flowering times of two plants the greater the probability of crossing. It seems likely that neighboring plants will overlap to a greater extent than plants from different subpopulations or populations. Differences in the flowering time of ecotypes have been observed in numerous species (e.g., *Gilia capitata*, Grant 1952; *Lamium amplexicaule*, Bernstrom 1952; *Geranium robertianum*, Böcher 1947; *Clarkia xantiana*, Moore and Lewis 1965; *Agrostis tenuis*, McNeilly and Antonovics 1968). If adjacent populations were differentiated with regard to flowering time, then distant pollen parents would be at a disadvantage.

Differential Viability and Vigor

Products of outcrossing typically have higher viability than products of selfing during seed development and as established plants (Allard et al. 1968; Stern and Roche 1974; Janossy and Lupton 1976). Among wild plants, the liability of self-fertilization is best understood among conifers. The incidence of seed abortion (Koski 1971, 1973; Bishir and Pepper 1977), and defective or slow growing seedlings (Franklin 1970; Koski 1973) is much higher with selfing than with outcrossing. The genetic load per zygote averages more than eight lethal equivalents in some species of *Pinus* and *Pseudotsuga* (Franklin 1972; Sorensen 1969; Koski 1973). As a consequence of their mating system, zygote populations of conifers are often rather inbred; mature stands of trees are much less inbred than zygote or seed populations from which they were derived. Mean interparent distance thus will increase through the life cycle. Direct evidence for this shift is seen in the population of *Eucalyptus pauciflora*, where the average outcrossing rate for plants as seeds was 63 percent compared with 76 percent for plants as seedlings (Phillips and Brown 1977). If cohorts could be followed in natural populations, the outcrossing rate undoubtedly would continue to increase, for the mean heterozygosity in adults is much higher than in seeds or seedlings. Different outcrossing rates are seen even among seeds stored for different periods. In *Eucalyptus delegatus*, outcrossing rates in old seeds was 85 percent versus 66 percent in the most recent collection (Moran and Brown 1980). Heterozygosity also is higher in the adult phase of the life cycle than in the seed phase, suggesting differential viability after establishment.

In predominantly outcrossing crop species, plants of inbred lines usually are weaker (McWilliam and Griffing 1965; McWilliam et al. 1969; Pederson 1968) and have a narrower ecological amplitude (Adams and Shank 1959; Jones 1958; Bucio Alanis and Hill 1966) than intervarietal hybrids. Similar differences may be expected in natural populations. Thus if neighboring plants are related, crosses between them might yield lower quality progeny than crosses between distant plants.

Some differences among progeny as a function of interparent distance are known. One component of seedling size, epicotyl length, was studied in the seeds of crosses of *Picea glauca*. Epicotyls of progeny of closely spaced parents were slightly shorter than those from distantly spaced parents (Coles and Fowler 1976). These data suggest that nearby parents were related when one considers that the epicotyls of self-pollinated seedlings were 24 percent shorter than those from long-distance outcrosses. Differences in seedling size have obvious fitness implications in that the larger seedlings within a cohort usually have higher survivorship than the smaller ones (Harper 1977).

Price and Wasser (1979) ascertained the effects of interparent distance on seedling survivorship in *Delphinium nelsoni*. They found that seedlings from crosses between plants 10 m apart had higher survivorship in the native sites than seedlings from crosses between plants 1 m, 100 m, and 1000 m apart, respectively. Self-seedlings had the lowest survivorship. We may conclude that seedlings from crosses between neighboring plants suffered inbreeding depression. Price and Wasser suggest that the progeny of wide crosses suffered outbreeding depression as a consequence of population subdivision and differentiation. The extent to which microdifferentiation played a role in this pattern of survivorship remains to be determined.

Survivorship which is a positive function of interparent distance has an interesting implication. As the season progresses, and genotype-dependent germination and mortality take their toll, the survivors will have parents which are, on the average, more widely spaced than the parents of plants that died. Correlatively adult plants will have parents which are on the average more widely spaced than the parents of immature plants. Mean parental distances thus increase through time!

As the distance between two plants increases, their genetic similarity is apt to decline, especially if the plants are not members of the same population (Levin and Kerster 1974). Given that plants from different populations will be more genetically dissimilar than plants from the same population, does that mean that distant plants are superior pollen parents vis à vis a given egg parent? In predominantly outbreeding crop species, the vigor and fecundity of hybrids increases as the genetic distance between parental strains increases until some critical level of divergence is reached when interactions at a few loci counterbalance the effect of heterozygosity. In maize, heterosis is a positive function of the level of divergence of strains except for the most divergent ones (Moll et al. 1965). In *Nicotiana* and *Gossypium* (Marani and Avieli 1973) heterosis also increases with genetic distance up to a point (Matzinger and Wernsman 1967).

In natural population systems, maximum heterosis is also associated with moderate levels of divergence. Heterosis has been documented in interracial crosses of Norway spruce (Nilsson 1974), Douglas fir (Orr-Ewing 1969), and loblolly pine (Woessner 1972; Owino and Zobel 1977). Heterosis in *Mimulus* is best developed in hybrids between populations which have undergone moderate degrees of divergence regardless of their geographical relationships (Vickery 1978). Beltran and James (1974) demonstrated heterosis in hybrids between chromosomally homozygous populations of *Isotoma petraea;* the more inbred the populations the greater the vigor of their hybrids. No correlation was found between heterosis and the level of population divergence. The *Isotoma* study is of particular interest because it shows that the genetic structure of populations, as shaped by population size and breeding structure, affects the relative quality of distant pollen parents.

Although heterosis is associated with moderate levels of differentiation, hybrids between differentiated populations are not necessarily heterotic. Indeed, they may be inferior to the products of intrapopulation crosses. For example, hybrids between some varieties of *Oryza sativa* are chlorotic by virtue of the action of two complementary dominant lethals (Oka 1957). Hybrids between *Oryza breviligulata* (wild) and *O. glaberrima* (cultivated rice) frequently are weak, owing to the presence of two complementary dominant genes (Chu and Oka 1972). Two aberrant developmental syndromes have been described in intervarietal hybrids of hexaploid *Triticum*. One is referred to as hybrid necrosis, which is controlled by two complementary dominant genes (Hermesen 1963). The other is referred to as grass-clump dwarfness, which results from the activity of three complementary dominant genes (Canvin and McVetty 1976).

Our knowledge of interpopulation hybrids is based principally upon crosses between populations ten to hundreds of miles apart, and upon performance trials in greenhouses, gardens, or plantations. Of principal concern here are hybrids between neighboring populations or those within pollination range, and how those hybrids would fare in natural populations. Neighboring populations may be well differentiated as a result of selective differentials or random drift. If populations are different by virtue of divergent adaptations, then interpopulation hybrids may be ill-fit in either population; it is unlikely they will be better adapted than local residents. On the other hand, if populations have diverged as a result of stochastic processes and are genetically depauperate, then interpopulation hybrids may be superior (heterotic) to local residents in both the egg and pollen parent populations. Under these circumstances, distant pollen parents would be superior to local pollen parents.

Interpopulation Hybridization

The percentage of plants which are interpopulation hybrids is poorly understood because of the difficulty in identifying them. The quantification of interpopulational hybridization requires substantial interpopulation differences in at least one readily assayed genetic marker. Ideally, populations would be homozygous, but differ in alleles, so that all interpopulation hybrids would be evident. Neither this nor the previous condition are likely to be found in nature.

Plant breeders have long been concerned with distances sufficient to preclude interpopulation hybridization. In many outcrossing crops, more than 5 percent of the seeds produced in one population have their pollen parent in another population, when the distance between populations is less than 100 m (cf. Kernick 1961). The incidence of interpopulation hybridization also is dependent upon the numbers of plants within a population. The smaller the population relative to those in the immediate vicinity, the higher the incidence of interpopulation hybrid seed (Fryxell 1956; Bateman 1947; Crane and Mather 1943; Williams and Evans 1935).

To the extent that interpopulation hybridization occurs, most parental pairs probably reside in adjacent populations. This is suggested by the restricted distribution of most animal- and wind-borne pollens (Free 1970; Levin and Kerster 1974). It is unlikely that pollen of many species is regularly transported more than 10 km, so that all but the closest populations may be out of the pollination range of a given population. Furthermore, as the pollen cloud moves away from a population it becomes progressively diffuse so that the ratio of indigenous to extraneous pollen reaching flowers will increase with distance from a source population. If so-called distant pollen parents reside in an adjacent population, then we would not expect their genetic constitution to be markedly divergent from the recipient population by virtue of diversifying selection. However, if populations were small, differentiation may have ensued from genetic drift. It would seem that whatever advantage distant pollen parents might have (in terms of crossability or progeny quality) it would be small—less than that described in crosses between distinctive plant varieties or distant plant populations.

In the absence of genetic markers, long-distance pollination can be demonstrated in nature by using single plant isolates. If a self-incompatible plant located between two populations sets seed, this is evidence that pollen can be effectively transmitted at least the distance between the isolate and the nearest population.

The higher the seed-set, the greater the flow of pollen from the populations, and perhaps between them. In the spring of 1979, I located several isolates of the self-incompatible, annual *Phlox drummondii* occurring 50–100 m away from the nearest population in areas where *Phlox* was common. On the average, 8 percent of the eggs of these plants were fertilized. Since pollinators (lepidopterans) bring pollen to isolates, they carry some away, perhaps depositing it within a nearby population. If there is pollen flow between isolates and neighboring populations, there probably is also pollen flow between populations tens and perhaps hundreds of meters apart.

Conclusions

Interparent distance provides another vantage point for analyzing the effective breeding structure of populations, a vantage point which integrates adaptive differentials throughout the life cycle. Given the dependence of pollen success and sporophyte survivorship on the genotype of parents and the likelihood of inbreeding in nature, cross- and self-compatible species probably are considerably more outbred than we might judge from pollination patterns alone.

Distance-dependent pollen success, and the seed and adult viability are possible in outcrossing populations only if seed dispersal is narrow, because this profile promotes the congregation of relatives in space. Then typical (viz. leptokurtic) pollen dispersal profiles result in breeding between relatives. In populations where seeds are broadcast over sizable distances, relatedness is not apt to be a function of distance, and local pollen dispersal will not foster inbreeding. As a consequence, the quality of pollen parents will not be dependent on distance, and the distribution of parent distances will be in accordance with the distribution of pollen dispersal distances.

Some implications of proximity dependent pollen parent success are evident within the context of Wright's neighborhood or isolation-by-distance model. A neighborhood may be defined as "an area from which the parents of central individuals may be treated as if drawn at random" (Wright 1969). The effective size of a neighborhood is equivalent to twice the number of reproducing individuals in a circle whose radius is equivalent to twice the standard deviation of the gene dispersal distance. A circle of this type will include 86.5% of the parents of the individuals at its center. The effective size and area of a neighborhood is dictated by the spatial relationships of parents and their offspring. For a population where seed dispersal is negligible, neighborhood size may be defined at $6.28\ \sigma^2\ d$ where σ^2 refers to the variance of pollen dispersal and d the density of flowering plants.

If the success of pollen and progeny increase with the distance between pollen and egg parents, then the variance of pollen dispersal does not describe the spatial relationships of the parents of adult plants. The variance of interparent distances will be much greater than the pollen dispersal variance; and when entered into the neighborhood formula, neighborhood size and area will increase. Accordingly, when pollen-parent success increases as the distance to egg parents increases, the likelihood of adult population subdivision by genetic drift declines, as does the degree to which two plants within a population are isolated by distance.

Neighborhood size and area may be estimated from gene flow profiles or the distribution of interparent distances as they are manifested in the zygote phase of the life cycle. It is tacitly assumed that neighborhood characteristics do not vary from one phase of the life cycle to another, so that the size and area of adult and zygote populations are equivalent. It is evident that the ranks of a cohort are substantially diminished during the sojourn through the life cycle. If seed, seedling, and immature plant mortality were independent of genotype, neighborhood characteristics would not change throughout the life cycle. However, if mortality were genotype-dependent—positively correlated with homozygosity—then the variance of parental distance would increase as the life cycle progressed; and the neighborhood size and area of the population would increase through the life cycle, reaching their peaks with the reproductive phase.

Finally, what are actual interparent distances for individuals in some population? Paterniani and Short (1974) came closest to answering this question. They established populations of maize in which a central plant was homozygous for a marker absent in the remaining plants. Progeny tests indicated the effective gene dispersal distance from single plants, from which they estimated the distribution of pollen parents and their numbers. Nearly 50 percent of the kernels of any individual plant result from pollen parents within a radius of 12 m. Several hundred plants were located within this radius. About 50 percent of the pollen parents were more than 12 m away. Thus the breeding structure in these fields approaches panmixia. This is somewhat counterintuitive if one considers only gene flow distances from the central plants. The gene dispersal curve in the aforementioned populations is highly leptokurtic with only a very small proportion of genes being dispersed beyond 6 m. Clearly leptokurtosis and narrow gene flow do not necessarily result in a tight breeding structure!

REFERENCES

Adams, M. W. and D. B. Shank. 1959. The relationship of heterozygosity and homeostasis in maize hybrids. *Genetics* 44:777-786.
Allard, R. W., S. K. Jain, and P. L. Workman. 1968. The genetics of inbreeding species. *Adv. Genet.* 14:55-131.
Allard, R. W. and A. L. Kahler. 1971. Allozyme polymorphisms in plant populations. *Stadler Symp.* 3:9-24.
Banyard, B. J. and S. H. James. 1979. Biosystematic studies in the *Stylidium crassifolium* species complex (Stylidae). *Aust. J. Bot.* 27:27-37.
Bateman, A. J. 1947. Contamination of seed crops. I. Insect pollination. *J. Genet.* 48:257-275.
Beltran, I. C. and S. H. James. 1974. Complex hybridity in *Isotoma petraea*. IV. Heterosis in interpopulational hybrids. *Aust. J. Bot.* 22:251-264.
Bernstrom, P. 1952. Cytogenetic intraspecific studies in *Lamium*. I. *Hereditas.* 38:163-220.
Bishir, J. and W. D. Pepper. 1977. Estimation of number of embryonic lethal alleles in conifers. I. Self-pollinated seed. *Silvae Genet.* 25:50-54.
Böcher, T. W. 1947. Cytogenetic and biological studies in *Geranium robertianum* L. *K. danske Vidensk. Selsk. Skr. Biol. Medd.* 20:1-29.
Brown, A. H. D., D. R. Marshall, and L. Albrecht. 1975. Estimation of

the mating system of *Eucalyptus obliqua* L. Herit. using allozyme polymorphisms. *Aust. J. Bot.* 23:931-949.

Bucio Alanis, L., and J. Hill. 1966. Environmental and genotype-environmental components of variability. III. Heterozygotes. *Heredity* 21: 399-407.

Canvin, D. T., and P. B. E. McVetty. 1976. Hybrid grass-clump dwarfness in wheat: physiology and genetics. *Euphytica* 25:471-483.

Chu, T. N., and H. I. Oka. 1972. The distribution and effects of genes causing F_1 weakness in *Oryza breviligulata* and *O. glaberrima*. *Genetics* 70:163-173.

Clegg, M. T., A. L. Kahler, and R. W. Allard. 1978. Estimation of life cycle components of selection in an experimental plant population. *Genetics* 89:765-792.

Coles, J. F. and D. P. Fowler. 1976. Inbreeding in neighboring trees in two white spruce populations. *Silvae Genet.* 25:29-34.

Crane, M. B., and K. Mather. 1943. The natural cross-pollination of crop plants with particular reference to the radish. *J. Apl. Biol.* 30: 301-308.

Crumpacker, D. W. 1967. Genetic loads in maize (*Zea mays* L.) and other cross-fertilized plants and animals. *Evol. Biol.* 1:306-424.

Dayton, D. E. 1974. Overcoming self-incompatibility in apple with killed compatible pollen. *J. Amer. Soc. Hort. Sci.* 99:190-192.

Frankel, R., and E. Galun. 1977. *Pollination Mechanisms, Reproduction and Plant Breeding*. New York: Springer-Verlag.

Franklin, E. C. 1970. Survey of mutant forms and inbreeding depression in species of the family Pinaceae. Southeast Forest Expt. Sta. USDA For. Serv. Res. Pap. SE-61.

Franklin, E. C. 1972. Genetic load in loblolly pine. *Am. Nat.* 106:262-265.

Free, J. B. 1970. *Insect Pollination of Crops*. New York: Academic Press.

Fryxell, P. A. 1956. Effect of varietal mass on the percentage of outcrossing in *Gossypium hirsutum*. *J. Hered.* 57:299-301.

Gorla, M. S., E. Ottaviano, and D. Faini. 1975. Genetic variability of gametophyte growth rate in maize. *Theor. Appl. Genet.* 46:289-294.

Grant, V. 1952. Genetic and taxonomic studies in *Gilia*. II. *Gilia capitata abrantifolia*. *Aliso* 2:363-373.

Harper, J. L. 1977. *Population Biology of Plants*. London: Academic Press.

Hermesen, J. G. T. 1963. The genetic basis of hybrid necrosis in wheat. *Genetica* 33:245-287.

Janossy, A., and F. G. H. Lupton eds. 1976. *Heterosis in Plant Breeding*. New York: Elsevier.

Johnson, C. M., and D. L. Mulcahy. 1978. Male gametophyte in maize: II. Pollen vigor in inbred plants. *Theor. App. Genet.* 51:211-215.

Jones, D. F. 1958. Heterosis and homeostasis in evolution and in applied genetics. *Am. Nat.* 92:321-328.

Kernick, M. D. 1961. Seed production of specific crops. *In* Agricultural and Horticultural Seeds, pp. 181-547. FAO Agricultural Studies No. 55.

Koski, V. 1971. Embryonic lethals of *Picea abies* and *Pinus sylvestris*. *Comm. Inst. For. Fenn.* 75:1-30.

Koski, V. 1973. On self-pollination, genetic load, and subsequent inbreeding in some conifers. *Comm. Inst. For. Fenn.* 78:1-42.

Levin, D. A. 1977. The organization of genetic diversity in *Phlox drummondii*. *Evolution* 31:477-494.

Levin, D. A. 1978. The origin of isolating mechanisms in flowering plants. *Evol. Biol.* 11:185-317.

Levin, D. A. 1981. Dispersal versus gene flow in plants. *Ann. Missouri Bot. Gard.* 68:233-253.

Levin, D. A., and H. W. Kerster. 1974. Gene flow in seed plants. *Evol. Biol.* 7:139-220.

Marani, A., and E. Avieli. 1973. Heterosis during the early phases of growth in intraspecific and interspecific crosses of cotton. *Crop. Sci.* 13:15-18.

Matzinger, D. F., and E. A. Wernsman. 1967. Genetic diversity and heterosis in *Nicotiana*. I. Interspecific cross. *Der Züchter* 37:186-191.

McNeilly, T., and J. Antonovics. 1968. Evolution in closely adjacent populations IV. Barriers to gene flow. *Heredity* 23:205-218.

McWilliam, J. R., and B. Griffing. 1965. Temperature-dependent heterosis in maize. *Aust. J. Biol. Sci.* 18:569-583.

McWilliam, J. R., B. D. H. Latter, and M. J. Mathison 1969. Enhanced heterosis and stability of growth of an interspecific *Phalaris* hybrid at high temperature. *Aust. J. Biol. Sci.* 22:493-504.

Moll, R. H., J. H. Lonnquist, V. Fortuno, and E. C. Johnson, 1965. The relationship of heterosis and genetic divergence in maize. *Genetics* 52:139-144.

Moore, D. M., and H. Lewis. 1965. The evolution of self-pollination in *Clarkia xantiana*. *Evolution* 19:104-114.

Moran, G. F., and A. H. D. Brown. 1980. Temporal heterogeneity of outcrossing rates in alpine ash (*Eucalyptus delegatus* R. T. Bak). *Theor. Appl. Genet.* 57:101-105.

Murakami, K., M. Yamada, and K. Takayanagi. 1972. Selective fertil-

ization in maize. Advantage of pollen from F_1 plants in selective fertilization. *Japan J. Breeding* 22:203-208.
de Nettancourt, D. 1977. *Incompatibility in Angiosperms.* Berlin: Springer-Verlag.
Nilsson, B. D. 1974. Heterosis in an intraspecific hybridization experiment in Norway Spruce. Proc. Joint IUFRO meeting S.02.04 1-3.
Oka, H. I. 1957. Complementary lethal genes in rice. *Japan J. Genet.* 32:83-86.
Orr-Ewing, A. L. 1969. Racial crossing in Douglas fir. Proc. Working Group on Quantitative Genetics Section 22 IUFRO North Carolina State University, Raleigh.
Owino, F., and B. Zobel. 1977. Genotype X environment interaction and genotype stability in loblolly pine. III. Heterosis and heterosis X environment interaction. *Silvae Genet.* 26:114-116.
Pandey, K. K. 1977. Mentor pollen: Possible role of wall-held pollen growth substances in overcoming intra- and interspecific incompatibility. *Genetica* 47:219-229.
Pandey, K. K. 1979. Overcoming incompatibility and promoting genetic recombination in flowering plants. *New Zealand J. Bot.* 17:645-663.
Paterniani, E., and A. C. Short. 1974. Effective maize pollen dispersal in the field. *Euphytica* 23:129-134.
Pederson, D. G. 1968. Environmental stress, heterozygote advantage, and genotype-environment interaction in *Arabidopsis*. *Heredity* 23: 127-138.
Pfahler, P. L. 1965. Fertilization ability of maize pollen grains. I. Pollen sources. *Genetics* 52:513-520.
Pfahler, P. L. 1970. In vitro germination and pollen tube growth of maize (*Zea mays*) pollen. III. The effects of pollen genotype and pollen source vigor. *Can. J. Bot.* 48:111-115.
Phillips, M. A., and A. H. D. Brown. 1977. Mating system and hybridity in *Eucalyptus pauciflora*. *Aust. J. Biol. Sci.* 30:337-344.
Price, M. V., and N. M. Wasser. 1979. Pollen dispersal and optimal outcrossing in *Delphinium nelsoni*. *Nature* 277:294-297.
Schaal, B. A. 1980. Measurement of gene flow in *Lupinus texensis*. *Nature* 284:450-451.
Sorensen, F. 1969. Embryonic genetic load in coastal Douglas-fir, *Pseudotsuga menziesii* var. *menziesii*. *Am. Nat.* 103:389-398.
Stern, K., and L. Roche. 1974. *Genetics of Forest Ecosystems.* Berlin: Springer-Verlag.
Thomson, J. D., and R. C. Plowright. 1980. Pollen carryover, nectar

rewards, and pollinator behavior with special reference to *Diervilla lonicera*. *Oecologia* 46:68-74.
Vickery, R. K. 1978. Case studies in the evolution of species complexes in *Mimulus*. *Evol. Biol.* 11:405-507.
Williams, R. D.,and G. Evans. 1935. The efficiency of spatial isolation in maintaining the purity of red clover. *Welsh J. Agr.* 11:164-171.
Woessner, R. A. 1972. Crossing among loblolly pines indigenous to different areas as a means of genetic improvement. *Silvae Genet.* 21:35-39.
Wright, S. 1969. *Evolution and Genetics of Populations.* Vol. 2 Chicago: University of Chicago Press.

8 Achieving Synthesis in Population Biology

WYATT W. ANDERSON

Population biology was born as the union of population genetics and population ecology some thirty years ago. Levene's note of 1953, on "genetic equilibrium when more than one ecological niche is available," signaled the beginning of this effort. The "Syracuse Symposium" of 1967 (Lewontin 1968) marked the full development of this hope for a synthesis of population concepts as much as any event, and Levins' (1968) seminal book *Evolution in Changing Environments* became a cornerstone of the new approach. The older field of *genecology* (Heslop-Harrison 1964), with its emphasis on selective effects of the environment on plants, became a part of the new population biology but not its core. Population biologists focused on the relationship of genetic variation to environmental variation, to the regulation of density, and to interactions among genotypes and between environment and genotype—topics still at the heart of the field.

Population biology has advanced considerably since the fifties and sixties. In such areas as theories of coevolution and demographic genetics, and both theoretical and experimental studies of life history adaptations, progress in relating ecological and genetic concepts has been substantial. Books such as Roughgarden (1979) and Charlesworth (1980) represent this growth of the field. Yet I believe it is fair to say that the original promise of a single, unified biology of populations has yet to be realized, and that large parts of population genetics and population ecology remain untouched by the other. There has been time for a synthesis, for by comparison the modern synthesis of evolutionary theory was accomplished during the years 1930 to 1950. The goal of a comprehensive population biology seems as attractive now as it did twenty years ago. To achieve a full synthesis, we must recognize some of the pitfalls which lie in the way of developing it. Hence I propose for discussion several problems which appear to have impeded the synthesis of population concepts, along with comments on a few topics which I believe will be foci for research in the next few years. My choice of points is not meant to be inclusive, but is rather a reflection of my own thoughts on population biology.

What Genetics and What Ecology?

Genetics at the population level is rather well circumscribed, since this level is the highest one at which genetic processes are treated. Ecology ranges from studies of individuals, to populations, and finally to communities and ecosystems. At the highest levels of organization, population genetics probably plays at best a minor role in determining whatever changes occur. This reasoning should not, however, lead to the conclusion that genetic changes are unimportant, or as one of my ecologist colleagues put it once, "only fine tuning." Genetics plays its role at the population level, in the short term through the process of adaptation, and over the long term through evolution. Evolutionary changes are then in turn reflected at higher ecological levels of organization, for instance by the genetic evolution of grazing faunas in areas where grasslands develop.

It is the processes of birth and death, and the interaction among organisms and between them and their environment that give rise to the phenomena of population biology, both genetical and ecological. Natural selection, genetic drift and founder effect, population growth and its regulation, competition, and predation are some of the basic processes at this level which have been linked in theory and experiment. Insofar as these processes have a common base in the life history schedules of the population, they operate on commensurate time scales and can be treated as coupled mechanisms.

Sometimes the environment may drive population sizes through rapid, even violent changes, as Gill, Berven, and Mock (article 1) have demonstrated for amphibians in pond communities. Genetic changes probably have little to do with these oscillations in number, but certainly the environmental factors driving them will have consequences for the genetic composition of the populations, genetic drift for example, and these genetic changes may well outlive concomitant changes in abundance and distribution. For instance, Carson (1974) has proposed that "founder" events may underlie the rapid formation of some Hawaiian *Drosophila* species, and here the genetic consequences of a colonization event clearly have exerted a continuing influence on these organisms long after their recovery from ecological perturbation. On the other hand, ecological processes such as competition are sometimes slow enough for genetic changes to accompany them. For instance, Maguire and Porter (1977) have studied competition among Pacific corals and found it to be a gradual process of interactions based on larval colonization, overgrowth, and extracoelenteric digestion. Here a study of both genetics and ecology would be most instructive.

Genetic responses to environmental perturbations are often muted by the large fraction of phenotypic variation which is environmental rather than genetic, and sometimes by the way genotypes interact with their environment. Many genetic changes may consequently be slower than ecological ones, and Lewontin (1979) has gone so far as to suggest that the two sets of processes operate on different time scales. There are certainly situations where the time scales differ, but for the reasons given above it seems unlikely this will generally be the case.

Theory and Experiment

The fusion of concepts in population biology has been based largely on mathematical models which explore relationships between such quantities as levels of genetic and environmental variation. These models have been extremely useful in creating a structure for the field. The problem here is a lack of experimentation to accompany the theory, to test predictions from it, and to suggest corrections. Critical experiments on rather fundamental aspects of population biology remain to be carried out. Theory has outpaced experiment so far that the theory has been released from the selective force of experimental test, and as a consequence it has become too elaborate and even in places ornate. These remarks are not a denigration of models but a lament for the loss of balance between theory and experiment.

An example is Levene's (1953) model, which probably represents the goal of population biology as well as any. Levene showed that organisms living in several habitats, with selection differing in them, can under certain restrictions on the breeding system maintain genetic polymorphisms without a requirement of heterozygote advantage in every habitat. These results were a considerable relaxation of the conditions for balanced genetic polymorphisms known at that time, and they suggested a relationship between environmental and genetic diversity. This model has been the subject of many refinements and extensions, enough so that at least one review of "Levene models" has been published (Christiansen and Feldman 1975). Surprisingly, there have only been a few laboratory experiments in which the relationship between environmental heterogeneity and the maintenance of genetic variability has been measured (Powell 1971; McDonald and Ayala 1974; Powell and Wistrand 1978), and not one of them has incorporated the rather stringent ecological conditions of Levene's model. There is still no critical test to this basic, often-cited model.

Life history schedules of longevity and fecundity underlie the formation of age structure in populations, and over the past decade a theory of population genetics in age-structured populations has been developed (Charlesworth 1980). The theory has been helpful in understanding possible causes of gene frequency changes in such situations as cycling vole populations, but there have been rather few direct experimental tests. Roughgarden (1977), and Fenchel and Christiansen (1977), have made great progress with coevolutionary models for the mutual adaptation of several species. Here particularly, experimental evidence is needed.

Synthesis

Density-dependent selection relates the key processes of population regulation and natural selection. It is the subject of an extensive theoretical literature and a fairly sizable experimental one. The concepts of r and K selection are a part of this topic, and it was attempts to set up and test predictions from the theory (see Pianka 1974) which led to an abandonment of a sharp $r-K$ dichotomy. This case illustrates the role experiment should play in keeping theory realistic. Other aspects of density-dependent selection have not been so carefully tested. Again, it is the balance between theory and experiment which seems out of order, not the theory itself.

Models Tend To Be Oversimplified

It is probably unavoidable that models in population biology present a simplified picture of the processes involved. Early in the development of the field, Levins (1966) discussed the tradeoffs necessary among precision, generality, and realism in constructing models. The problem is that model-building has continued to focus on overly simple genetic and ecological descriptions of processes, and on combinations of them.

Genetic effects, for example, are often analyzed in terms of two alleles at a single locus. Recent analyses of multiallelic and multilocus models have revealed a surprising richness of equilibria and a behavior not easily extrapolated from the simpler case. Sickle cell hemoglobin is a case in point. Heterozygote advantage for carriers of the sickle cell allele is taught as a classic case of natural selection. Where three alleles are known (Hb^a, Hb^s, Hb^c) the situation suddenly becomes more complicated and heterozygous advantage no longer suffices to explain the balanced polymorphism (Cavalli-Sforza and Bodmer 1971).

The genetic components of population biology models have concentrated too much on balanced polymorphisms. This kind of model is simply not appropriate for thinking about multifactorial inheritance, which covers many of the characters most important in adaptation. Istock's (article 3) research on pitcher-plant mosquitoes is a good example of the value of studying polygenic systems. He has evidence that appreciable genetic variability for life history characters is maintained in natural populations, contrary to expectations from prevailing theory.

Similar questions can be raised about the staple ecological models for population regulation and competition which figure prominently in population biology. The existence of higher-order interactions not treated in the usual formulations (see, e.g., Wilbur 1972) is particularly worrisome.

The way genetics and ecology are coupled is crucial to the conceptual development of population biology. Sometimes simple genetics is joined with simple ecology to good effect, as in Levene's (1953) model, mentioned already, or in MacArthur's (1962) original treatment of selection among genotypes differing in their responses to population density. Less effective is the modification of a fairly complex model from one field by the introduction of a simple version of effects from the other. For example, the competition between two species may be extended to include genotypic differences in competitive ability, which will then result in natural selection which in turn affects the overall competitive ability of each species (Leon 1974). If a single locus and two alleles are utilized, the model is already a bit oversimplified, since competition is unlikely to rest on a single gene, or even a few. If the ecological model is then elaborated to include resource utilization functions and other trappings of modern population ecology, the degree of ecological complexity and realism far exceeds the degree of genetic realism. The resulting model pays lip service to genetics and ecology but is not very useful, because the effort required to treat the genetic parameters is hardly worthwhile in light of the lack of realism in defining the genetic system.

Focus on Equilibrium States

The usual procedure in developing a model is to state the assumptions, define the model, and then work out the equilibrium properties, such as the conditions for a balanced polymorphism or a stable age structure. Populations are unlikely to be in equilibrium, however. They are much more likely to be in a state of constant flux. The *dynamics* of processes such as selection and competition are more important than the *statics* of an equilibrium seldom attained. Equilibria are important only insofar as the behavior of populations near them reflects the behavior away from them. Stability analyses near equilibria may in particular fail to indicate biologically important behavior away from equilibrium when genetical and ecological processes are coupled, so that changes in one component may drive changes in the other. Density-dependent selection, in which genotypic fitnesses are simple logistic functions of population size, is an example. Genotype differences in carrying capacity may produce a shoulder or apparent equilibrium in population number which later deteriorates to a quite different level as gene frequencies change and a slow-growing (low r) but density-resistant (high K) genotype increases in frequency (Anderson and King 1970; King and Anderson 1971).

An experimental example comes from studies with *Drosophila pseudoobscura*. The chromosomal inversions in this species tie up large blocks of genes as single units—essentially as superalleles of a supergene. The inversions undergo powerful selection in the lab (Wright and Dobzhansky 1964) and in nature (Dobzhansky 1943; Anderson et al. 1979), and this case is often used as a textbook illustration of natural selection. In fact, many populations in nature never approach a stable equilibrium point, but show continued fluctuations over a range of frequencies twenty to thirty percentage points wide. The selection cannot be constant, but must be shifting continually to produce such oscillations. Frequency-dependent effects may well be involved, since the inversion frequencies do not wander at random but are constrained to a band of frequencies. The concept Prout (1968) has called *protected polymorphism* is useful here, because it takes account of some of these problems; it is the method used by Levene (1953) in his classic paper. Instead of worrying about equilibria, we ask what are the conditions under which alleles will be maintained.

Optimization Arguments

Optimization arguments have great appeal, for they provide a goal to various processes. Fisher's fundamental theorem of natural selection, Wright's fitness functions, the optimization of time and energy in ecology, evolutionary stable strategies, and Levins' fitness set strategies are examples. Most are based on an underlying belief that some quantity should be optimized during the course of a population process. Few have withstood careful scrutiny. For example, the average fitness of a population does not always increase under selection when linked genes are considered (Moran 1964). Levins' fitness set approach to the consequences of selection in changing environments provides only one of several sets of possible solutions, and some of the alternatives may be more nearly optimal (Templeton and Rothman 1974).

More recently, the idea that existing population structures are adaptive has been used to establish conditions for maintenance of some feature, such as a particular life history schedule. This is the evolutionary stable strategy of Maynard Smith (1974). Such analyses can be useful, but the basic assumption that the existing state of affairs is optimal ought to be examined critically for each case, in order to avoid what Gould and Lewontin (1979) have called the "Panglossian paradigm." Much of the recent ferment in sociobiology and behavioral ecology is based on the premise of optimality. Can it be tested? Can fitness measurements be made of behaviors departing from the assumed equilibria?

Bringing Behavior Fully Into Population Biology

Sociobiology and behavioral ecology have arisen as fields since the mid-sixties, when the idea of combining population genetics and population ecology was at its peak. The role of behavior in the biology of animal populations is clearly a major one, and behavior must be made a partner with ecology and genetics in any true synthesis of population concepts.

Hamilton's (1964) theory of inclusive fitness opened the way to understanding how social behaviors could evolve, and it became the cornerstone of sociobiology. There are as yet few tests of this theory, however, and some of its most basic assumptions remain to be firmly established.

For instance, the assumption that behavioral patterns are adaptive is a cornerstone of behavioral ecology and sociobiology; this assumption needs rigorous testing. Another basic assumption of sociobiology is that genetic differences exist for complex forms of behavior such as altruism, selfishness, and other social acts; it remains undemonstrated. Genes producing variant behaviors are known in many organisms, and in some cases the physiological basis for the behaviors are known. But these behavior patterns are not like those involved in sociobiological models, and here again the assumptions warrant critical evaluation.

Sociobiological theory has grown rapidly and perhaps it is no surprise that some of it is uncritical. Sorting out the theory and testing it are major goals of the next decade. I think sociobiology is, in Lakatos' (1970) words, a progressive research program, that the best measure of its success is the new ideas and hypotheses it is generating and the ferment they are causing in other areas of population biology.

Connecting Phenotypes to Gene Function

Population processes such as selection act on phenotypes, and it is essential to connect phenotypes to the genetic units which produce them. Population processes will be understood only when they can be explained on the basis of the organism's physiology and the biochemistry of gene action. Then the way an organism adapts to its environment, and the way its genotype acts in doing so, can be explained and even predicted. For a few years there was a great deal of discussion about what the operational genetic units are—whether, for instance, they are genes or blocks of genes held together by linkage (Franklin and Lewontin 1970). Theoretical studies proved a poor guide here, for measurements in natural populations failed to show much linkage disequilibrium except for those situations where the breeding system or cytology produced it.

Clarke (1975) appears to have been correct in urging case-by-case studies of connections between the proteins encoded by loci and the physiology of the organism which brings about adaptation to the environment. Only a few studies have been reported, and this is an area which will surely have a significant impact on population biology in the next decade. Koehn (article 5) has discussed his research, which is one of the best examples of this approach. Another is the work of Watt and his colleagues (Watt 1973, Graham et al. 1980) on *Colias* butterflies. They have made detailed analyses of the sex-limited "alba" color polymorphism, caused by a dominant mutant at a single locus. In the pupal stage of alba females, nitrogen-rich precursors are diverted from the synthesis of wing pigments, with the result that eclosion and maturation of eggs are faster. Fitness is thus increased as a result of changes in the mass-energy-time budget. A counteracting selection acts against alba females because the white ground color of their wings is less attractive to males than the normal color. Conflicting selection pressures, as in this case, are probably more the rule than the exception and serve to warn against simple solutions; the fitness of the alba phenotype will depend on both physical and biological aspects of the environment. Conflicting selection pressures at different life stages do not produce a balanced genetic polymorphism, of course. It is the net selection on the genotypes which determines whether or not a polymorphism will be maintained.

It is necessary to ask whether studies like this, or of the Adh locus polymorphism in *Drosophila melanogaster*, or of Koehn's studies with *Mytilus*, are typical. They involve genes coding for enzymes with a direct effect on the physiology of the organism whose adaptation is under study. Many other loci may be involved in minor reactions or in the steps of an alternative metabolic pathway, so that the effects of genetic change may be to produce minor adjustments in the flow of metabolites through a pathway. Perhaps these adjustments are to be expected for polygenic variability, characterized as it is by genes with small individual effects. Understanding how such genes affect the phenotype on a physiological level will be difficult, but nonetheless important.

Methodology of Research

Population genetics is an unusual area of biology in that a mathematical structure for the field was developed before experimentation could test its assumptions and hypotheses derived from the theory. Population ecology also had an early theoretical base, which was partially submerged in a long period of descriptive research. More recently, such workers as MacArthur and May have developed a new ecological theory. A theory, or at least *components* of a theory, of population biology now exists. This theory poses difficulties for the experimenter, for it is often not precise enough to generate testable hypotheses. Usually it is more heuristic, its value being in the classes of relationships it predicts or the processes which it suggests for observation and experiment. The theory is too general, however, and it needs critical assessment.

In recent years there has been an emphasis on introducing stricter standards of testing hypotheses in population biology. Ayala (e.g., Ayala et al. 1975) has been particularly influential in introducing Popper's (1959) ideas of scientific methodology as a series of falsifications of hypotheses. This methodology has a salutary effect in forcing the scientist to state just what the purpose of an experiment was, to decide whether the results refute the hypotheses, and to consider whether there was a reasonable possibility they could have done so. There is nothing really new in this approach, but it is surprising how many studies ignore such basic considerations. Here population biologists can take a lesson from molecular biologists; the sort of "strong inference" which Platt (1964) used to characterize rapid advances in molecular biology is, I believe, more difficult to employ in population biology but equally worthwhile. How often does a paper in our field list all major alternative hypotheses and describe how they were tested and why they were rejected? The way in which Connor and Simberloff (1978) and Strong, Szyska, and Simberloff (1979) test ecological hypotheses against null alternatives illustrates the power of this approach.

Importance of Technical Advances and a Direct Approach

My colleagues in molecular genetics have impressed on me the importance of new techniques in answering stubborn research problems. Certainly the dramatic success of the recombinant DNA technology in providing the means to answer questions about gene organization and function bears testimony to this point. The new perspective on genic variation which opened up in the mid-sixties was the result of a new technique, electrophoresis, borrowed from molecular biology. An essential feature of advances in molecular biology has been the continual search for new means of *directly* testing hypothesis—that is, for physical manipulation or control of a system. Population biology has relied too heavily on *inferential* approaches to questions, based on analysis of data within the context of a model, so that the data themselves subtly take on certain features of the model which may not actually exist. The analysis of selection is a good example. Early studies were simply observations of gene frequency changes, with fitnesses estimated from the theory, which at that time assumed selection of constant intensity (Wright and Dobzhansky 1946). Only in the past decade or so have direct measurements of fitness components been obtained at various life history stages: in *Drosophila*, for example, larval viability, female fecundity, and male mating success (Prout 1971). As simple as it sounds, the results of this direct approach to measuring selection were surprising; one finding was that the fertility component—female fecundity and male mating success—was often more important than viability in bringing about gene frequency changes. In this case, no really new techniques were required, but rather a new attitude toward problems.

In other cases, technical advances will govern the ability to solve problems. The introduction of restriction enzyme techniques to the question of measuring genetic relationships is an example. Avise et al. (1979) have studied the similarities of mitochondrial DNA in mice as an index of their evolutionary relatedess. Since the mitochondria are transmitted maternally, this technique allows the delineation of individuals descended from a common mother sometime in the past; this information is particularly helpful in close relationships.

Summary

Population biology seems to have stopped short of the synthesis of genetical and ecological concepts it promised twenty years ago. I have briefly considered several problems I believe have impeded a full synthesis. The first involves the levels at which genetics and ecology can be combined and the matter of time scales, and I conclude that the processes rooted in birth and death schedules and in interactions within and between populations, which form the core of population biology, generally operate on commensurate time scales. The most serious problem I see is an imbalance between theory and experiment which has led to a very rich framework of mathematical models, most of which have not been critically tested and corrected in the light of the experimental evidence. Instead the theory has grown on itself in such a way that what has been proven is sometimes difficult to distinguish from what has been conjectured. The models which constitute this theory tend to be oversimplified descriptions of populations. Sometimes they fail to connect basic genetical and ecological processes at their roots but are, rather, embellishments of models from one field with a few aspects of the other. Theory is focused too much toward equilibrial states, when in fact many populations never attain equilibria but are governed for long stretches of time by transient behavior. Optimization arguments are another problem because their intrinsic appeal may direct attention away from rigorous examination of the assumptions underlying them. A critical methodology based on strict hypothesis testing and on full scrutiny of alternative hypotheses must be employed if the theoretical superstructure of population biology is to be molded into fit with nature. I stress the importance of direct experimental tests of hypotheses over short steps of reasoning, as contrasted with inference over long chains of concepts, as well as the devising of new experimental techniques to answer questions which seem out of the reach of experimentation now. I comment on two topics at opposite ends of the biological spectrum as necessary ingredients in a comprehensive biology of populations; at the organismic end, behavior, despite all the problems which currently affect sociobiology and behavioral ecology; and, at the molecular end, the relationship of phenotypes in nature to the underlying physiology and biochemistry of the organism and its genes.

ACKNOWLEDGMENTS

I am grateful to my fellow participants in the symposium, and to John Avise, Jerry Coyne, and Tim Prout for critical comments which helped me in revising this paper.

REFERENCES

Anderson, W. W. and C. E. King. 1970. Age-specific selection. *Proc. Nat. Acad. Sci. USA* 66:780-786.

Anderson, W. W., L. Levine, O. Olvera, J. R. Powell, M. E. de la Rosa, V. M. Salceda, M. I. Gaso, and J. Guzman. 1979. Evidence for selection by male mating success in natural populations of *Drosophila pseudoobscura*. *Proc. Nat. Acad. Sci. USA* 76:1519-1523.

Avise, J. C., R. A. Lansman, and R. O. Shade. 1979. The use of restriction endonucleases to measure mitochondrial DNA sequence relatedness in natural populations. I. Population structure and evolution in the genus *Peromyscus*. *Genetics* 92:279-295.

Ayala, F. J., J. W. Valentine, T. E. Delaca, and G. S. Zumwalt. 1975. Genetic variability of the antarctic brachiopod *Liothryella notorcadensis* and its bearing on mass extinction hypothesis. *J. Paleontol.* 49:1-9.

Carson, H. L. 1974. Patterns of speciation in Hawaiian Drosophila inferred from ancient chromosomal polymorphism. In M. J. D. White, ed, *Genetic Mechanisms of Speciation in Insects*. Sidney: Australia and New Zealand Book Co.

Cavalli-Sforza, L. L. and W. F. Bodmer. 1971. *The Genetics of Human Populations*. San Francisco: W. H. Freeman.

Charlesworth, B. 1980. *Evolution in Age-Structured Populations*. Cambridge: Cambridge University Press.

Christiansen, F. B. and M. W. Feldman. 1975. Subdivided populations: A review of the one- and two-locus deterministic theory. *Theoret. Pop.* 7:13-38.

Clarke, B. 1975. The contribution of ecological genetics to evolutionary theory: Detecting the direct effects of natural selection on particular polymorphic loci. *Genetics* 79:101-113.

Connor, E. F. and D. S. Simberloff. 1978. Species number and composi-

tional similarity of the Galapagos flora and avifauna. *Ecol. Monographs* 48:219-248.
Dobzhansky, Th. 1943. Genetics of natural populations. IX. Temporal changes in the composition of populations of *Drosophila pseudoobscura*. *Genetics* 28:162-186.
Fenchel, T. and F. B. Christiansen. 1977. Selection and interspecific competition. In F. B. Christiansen and T. M. Fenchel, eds. *Measuring Selection in Natural Populations*. Heidelberg: Springer-Verlag.
Franklin, I. and R. C. Lewontin. 1970. Is the gene the unit of selection? *Genetics* 65:707-734.
Gould, S. J. and R. C. Lewontin. 1979. The spandrels of San Marcos and the Panglossian paradigm: A critique of the adaptationist programme. *Proc. Roy. Soc. London, Ser. B.*, 205:581-598.
Graham, S. M., W. B. Watt, and L. F. Gall. 1980. Metabolic resource allocation vs. mating attractiveness: Adaptive pressures in the "alba" polymorphism of Colias butterflies. *Proc. Natl. Acad. Sci. USA* (in press).
Hamilton, W. D. 1964. The genetical evolution of social behavior I and II. *J. Theoret. Biol.* 7:1-52.
Heslop-Harrison, J. 1964. Forty years of genecology. *Adv. Ecol. Res.* 2:159-247.
King, C. E. and W. W. Anderson. 1971. Age-specific selection. II. The interaction between r and k during population growth. *Am. Nat.* 105: 137-156.
Lakatos, I. 1970. Falsification and the methodology of scientific research programmes. In I. Lakatos and A. Musgrave, eds. *Criticism and the Growth of Knowledge*. Cambridge: Cambridge University Press.
Leon, J. A. 1974. Selection in contexts of interspecific competition. *Am. Nat.* 108:739-757.
Levene, H. 1953. Genetic equilibrium when more than one ecological niche is available. *Am. Nat.* 87:311-313.
Levins, R. 1966. The strategy of model building in population biology. *Am. Sci.* 54:421-431.
Levins, R. 1968. *Evolution in Changing Environments*. Princeton: Princeton University Press.
Lewontin, R. C., ed. 1968. *Population Biology and Evolution*. Syracuse: Syracuse University Press.
Lewontin, R. C. 1979. Fitness, survival and optimality. In D. J. Horn, R. D. Mitchell, and G. R. Stairs, eds. *Analysis of Ecological Systems*. Columbus: Ohio State University Press.
MacArthur, R. H. 1962. Some generalized theorems of natural selection. *Proc. Nat. Acad. Sci. USA* 48:1893-1897.

McDonald, J. F. and F. J. Ayala. 1974. Genetic response to environmental heterogeneity. *Nature* 250:572-574.

Maguire, L. A. and J. W. Porter. 1977. A spatial model of growth and competition strategies in coral communities. *Ecological Modeling* 3:249-271.

Maynard Smith, J. 1974. The theory of games and the evolution of animal conflicts. *J. Theoret. Biol.* 47:209-221.

Moran, P. A. P. 1964. On the nonexistence of adaptive topographics. *Ann. Human Genetics* 27:383-393.

Pianka, E. R. 1974. *Evolutionary Ecology*. New York: Harper and Row.

Platt, J. R. 1964. Strong inference. *Science* 146:347-353.

Popper, K. 1959. *The Logic of Scientific Discovery*. New York: Harper.

Powell, J. R. 1971. Genetic polymorphism in varied environment. *Science* 174:1035-1036.

Powell, J. R. and H. E. Wistrand. 1978. The effect of heterogeneous environments and a competitor on genic variation in *Drosophila*. *Am. Nat.* 112:935-947.

Prout, T. 1968. Sufficient conditions for multiple niche polymorphism. *Am. Nat.* 102:493-496.

Prout, T. 1971. The relation between fitness components and population prediction in *Drosophila* I and II. *Genetics* 68:151-167.

Roughgarden, J. 1977. Coevolution in ecological systems: Results from "loop analysis" for purely density-dependent coevolution. In F. B. Christiansen, and T. M. Fenchel, eds. *Measuring Selection in Natural Populations*. Berlin-Heidelberg-New York: Springer-Verlag.

Roughgarden, J. 1979. *Theory of Population Genetics and Evolutionary Ecology: An Introduction*. New York: MacMillan.

Strong, D. R., L. A. Szyska, and D. S. Simberloff. 1979. Test of community-wide character displacement against null hypotheses. *Evolution* 33:897-913.

Templeton, A. R. and E. D. Rothman. 1974. Evolution in heterogeneous environments. *Am. Nat.* 108:409-428.

Watt, W. B. 1973. Adaptive significance of pigment polymorphisms in Colias butterflies. II. Progress in the study of the "alba" variant. *Evolution* 27:537-548.

Wilbur, H. M. 1972. Competition, predation, and the stucture of the *Ambystona-Rana sylvatica* community. *Ecology* 53:3-21.

Wright, S. and Dobzhansky, Th. 1946. Genetics of natural populations. XII. Experimental reproduction of some of the changes caused by natural selection in certain populations of *Drosophila pseudoobscura*. *Genetics* 31:125-156.

Biology Colloquium
Roundtable Discussion

At the conclusion of the colloquium, participants gathered for a two-hour extemporaneous discussion on both the present status and future directions of population biology. The discussion was recorded and appears in condensed and edited form below. Editing involved both elimination of extraneous and peripheral comments and reorganization of some of the statements to achieve a greater coherency. Insofar as possible, the flavor of the interchanges has been preserved. Our introductory comments to each topic discussed appear in italics—eds.

I. Measuring Genetic Variation

One point repeatedly arose during the colloquium: Is there a continuing need to be concerned with measuring the amount of genetic variation? Appropriately, therefore, the Roundtable Moderator, Wyatt Anderson, asked Jeff Powell to start the discussion by presenting his views.

Powell. I think there are three reasons why we should be concerned with measuring genetic variation. The first is that I don't think we have sufficient knowledge of the levels of genetic diversity that need to be accounted for. Once we have adequate data, our choice of realistic models will be narrowed. For instance, if all gene loci were like XDH in *Drosophila pseudoobscura*, and had 20, 30, or 40 alleles, then as Lewontin et al., (*Genetics* [1978] 88:149-170) showed, heterosis is not a realistic model and should be discounted. If it were common for posttranslational modification of several loci with two and three alleles to generate a whole series of proteins, heterosis could be rescued as a viable model.

A second reason that lends importance to measuring genetic variation concerns our view of molecular evolution. If protein polymorphisms, or a significant proportion of them, represent an intermediate stage in the evolution of proteins, then it is important to know the number of protein-coding loci that may be undergoing the substitutional process at any given time. It might also help us understand problems such as substitutional load.

Finally, a third and perhaps more practical point, is to use protein polymorphism to study geographical differentiation and taxonomic differentiation. The classical picture derived from allozyme studies is that populations of any given species are relatively homogeneous. The same allele is more or less at the same frequency in all geographic populations; species display remarkable genetic homogeneity. We need to know whether the homogeneity is an artifact of electrophoresis and once we get beyond that are we going to see much more geographic differentiation? The same is true at higher taxonomic levels when we try to measure the amount of genetic divergence which occurs during the speciation process. How much difference is there really between subspecies and species? Allozyme data have already been useful for studying the genetic structure of populations, but we may need more sensitive techniques.

Koehn. Since I was the one who argued yesterday against continuing to measure genetic variation, I would like to reply. I think Jeff [Powell] is perfectly right in one sense; electrophoresis needs no defense in the sort of population biological studies that Don Levin described. The technique will probably always be used for studies of population structure. I also agree that it will be of continuing usefulness for examining patterns of taxonomic differentiation. We need to know more about the speciation process;

since we know so little about it, almost anything we discover will be a step in the right direction. We can use more data on patterns of geographic differentiation and population structure. For all of these reasons, electrophoresis was really a major technological breakthrough in population biology.

However, I don't think any of these questions depend directly upon an assessment of genetic variation. I do not agree with Jeff on his other points. Although molecular evolution is certainly concerned with how alleles are spatially or temporally substituted and/or maintained in populations, I disagree with the contention that refined measures of genetic variation might cause us to throw out heterosis as a model. I do not think we will throw out heterosis as a model no matter what we find. Such a statement is predicated on the assumption that there is but one model of selection that explains *all* variation, and that assumption is demonstrably untrue. We will continue to search for evidence of this mode of selection, as well as alternative modes. I think that we will come to the point where we recognize that different models apply in different cases.

King. To take the point a bit futher, the absolute level of genetic variation in any given population is going to be the end result of a number of different processes, and it seems to me hopelessly naïve to search for a single factor that will provide a holistic explanation. Instead, we should focus our studies on the processes themselves and how they interact to influence levels of genetic variation. In the absence of other vital infomation, the absolute level of genetic variation is simply another descriptive aspect of population structure.

Istock. I'd like to say something in defense of Powell. I think we've been a bit unfair and we are beating a dead horse. The young people that I know refer to the kind of thing that Powell has been criticized for, and doesn't do much of, as the "find 'em and grind' em" school of population biology. I prefer my own version which is the "smash 'em and shock'em" view. But I think it's a problem of a dead horse—it's a useless thing. I think we're beginning to mature now. We know that sometimes, for some problems, we should measure amounts of structural gene variation, but other times we can go on to other questions and methods.

II. Give Me a Place to Stand..."

The short history of population biology, and the much longer tradition of its antecedents, can be characterized as a search for the vantage point of Atlas. It is not that population biologists are so naïve as to think that a single character or a single analytical method can be used to describe all important aspects of population structure. Rather, it is dissatisfaction with the known limitations and flaws of the characters and methods currently available that leads to rapid utilization of new measures by such a large proportion of the scholars in the field. Two relatively recent examples of this "bandwagon" effect are the use of electrophoresis to characterize genetic variation, and the use of information-theory–based diversity measures to characterize community structure.

In any gathering of population biologists, one need not look too deeply before confronting the compound question of "what should we measure, and how should we measure it?" As indicated by the following interchange, the Colloquium at Oregon State University was no exception.

Levin. Jeff [Powell] was discussing the need to understand levels of genetic variation. I am interested in what the panel considers to be the characters that should be studied. Obviously some characters are less important than others. Are we looking at the right characters in terms of local adaptation? Does anyone care to comment on the characters that should be chosen, or in which directions we need to go?

Anderson. I think there is a movement back to studying morphological characters directly involved in adaptation. Locating the genetic variability for them is difficult.

Istock. But Don [Levin] did pretty well in his own work. Isn't it plausible to view every demographic event as potentially a selective event or a drift event, unless the individual is senescent? Whenever an individual does not complete its maximum reproductive life expectancy, but dies early or fails to reproduce maximally, we are observing potential selection. That's what we are measuring if we follow through the life cycle looking for failures and their net effect on individual reproductive rates.

Levin. Demography seems to integrate the processes, but doesn't tell you about the characters in question. My own impression is that plant evolutionists have been looking at morphological and allozymic characters for several years; perhaps that's not where the action is with regard to selection in natural populations. We know very little about variation in physiological processes, photosynthetic rates, or transpiration rates. Certainly metabolic pathways must vary between organisms. We know very little about developmental variables. Perhaps we've been looking at the wrong characters and the patterns of morphological and allozymic variation in space usually do not really speak to the issue of local adaptation.

Koehn. I'm not sure they're so different from one another. It is just that we don't understand how they may be interconnected, which is, no doubt, in a very complex way. The point I wanted to make is that I think our most important allies in population biology in the next decade will be the physiologists, particularly the ecological physiologists. I talked a bit about this yesterday with respect to a single gene. The reason I became interested in this area was to understand the selective processes that we monitor with the use of allozymes. I would like to know why an individual of one genotype dies in certain ecological circumstances when another one doesn't. I want to know when an animal is in trouble. I want to know the effects on an organism of experiencing specific ecological circumstances, particularly the effect on measures of fitness. What do I mean by that? We can measure such an effect directly from various physiological parameters. Measures of physiological stress, which presumably precedes death, may be manifest in different ways, such as changes in respiration rates, ability to mobilize energy reserves, and a variety of other measurements that are routinely made by ecological physiologists. These can tell us a lot about whether an organism is or is not under stress. If we can tie physiological characters to some genetic basis, we can perhaps start to get these different kinds of characters together. In a sense, I tried to do this yesterday by providing an example of how this may work in the case of a single gene. That is not to say I was talking about single gene evolution; it's going to be a lot more complex than that. However, I think the same general approach is valid.

Powell. In this context, I think our best friends are going to be the evolutionarily oriented molecular biologists. There is a real technological revolution occurring in molecular biology and we've got to convince our molecular colleagues that the kinds of problems we're grappling with are valid and important and that they should help us. Perhaps they could use their expertise to aid us in solving some of our problems such as how gene regulation and developmental processes might be related to fitness. Francis Crick (*Science* [1979] 204:264-271) made the statement that any molecular biologist who's going to understand the evolution of genomes will have to know some modern ideas of population genetics. I would add the converse: the population geneticist who's going to really understand how genomes evolve is going to have to keep up with and use the latest developments in molecular biology.

Levin. Before we get away from the technology that Jeff [Powell] and Dick [Koehn] mentioned, it seems to me that we are not quite as dependent on molecular biologists as they suggested earlier. I think there are many problems that have relatively simple solutions if we would just sit and think about them. For much of population biology, you don't have to know the molecular basis of "X", "Y" and "Z" in order to demonstrate certain population phenomena. The experiment on gene dispersal I talked about earlier was quite simple. Just put some plants with a homozygous dominant marker in the field and collect seeds at various distances from these individuals. Plant the seeds, score the phenotypes, and you have concrete dispersal data. In many areas of population biology, I believe that transparent experiments made with just a little insight or common sense will provide solutions that sophisticated instruments or chemical analyses cannot. So I'm still plugging away for common sense and imagination.

Dawson. I think Wyatt [Anderson] made that point very well when he said all it takes is some fresh approaches. What Dick Levins did for our field didn't require any money, it just required thinking about the problem of adaptation to environmental heterogeneity in a novel way. And the same point is surely true for experimental attempts to test the theory of population biology.

Koehn. It's a mistake to naïvely equate progress with technology. There are questions that in fact do depend on technology, but you can ask interesting questions without technology or you can pursue uninteresting questions with technology. I was going to say very much the same thing that Jeff [Powell] said about the impact of molecular biology. It's quite obvious that the field is undergoing an explosion and it's going to have an influence on our thinking, as it already has in molecular evolution. All my molecular geneticist friends are urging me to go into nucleic acid sequencing. I'm certain Wyatt is having the same experience. The point is that it is a very exciting field; big discoveries are being made in molecular genetics and this excitement is spreading. My colleagues are saying "Why aren't you getting in on the ground floor, so to speak, in applying these techniques to population biology?" Frankly, I don't know how to apply them right now. I'm fully convinced, however, that this will come and I expect it; but right now I'm not sure how. Does anyone have any comments?

Anderson. Dick, I would have thought that you were the quintessential representative of the application of physiological-molecular techniques.

Koehn. Yes, probably, but not unless you've identified a specific question. Of course, I am thoroughly interested in the need you cited in your paper for connecting the phenotype to gene function. I will probably continue along those lines in the next ten years, because I think it is an important area. I think it is important from several perspectives; one is it would be nice to have a few model systems. We desperately need some good biochemical-ecological models. I don't know many such examples we need but we need some, at least. We need a few good examples of *mechanisms* of evolution. Another perspective of importance has to do with regulator genes and modifier genes. I think we sometimes lose sight of the fact that modifier genes modify structural genes, as for example in ADH of *Drosophila*. And, I would assume that because modifier genes are modifying structural genes that some insight into the evolution of gene systems will come from looking at the properties of structural genes. Even though there is not much evidence for linkage in natural populations, implicit in the idea that coadapted genotypes evolve is that there are phenotypic alternatives for individual genes; coadaptation involves organization among these phenotypic alternatives. I believe that much information on evolutionary mechanisms can be obtained by identifying alternative phenotypes in terms of gene function. A bigger question is "Which gene should we go after?" I mean, shall we just pick our favorite old enzyme or is there some rationale that would help shortcut what is going to be a very tedious task? I think there is such a rationale. With LAP, for example, when you see a 50 percent spatial change in frequency, it tells you something interesting is going on. On the other hand, you had better not try to relate just any environmental variable to one gene or to a group of genes if there is no biochemical basis for a relationship. So I think there is some information in the biochemical function of an enzyme that tells you what physiological processes it can participate in and thereby which environmental parameter could affect its function.

The last thing I would like to add is a side comment on funding. To the extent that molecular biology filters into population biology, the cost of research increases phenomenally. I think that funding of population biology by NSF, etc, has got to be cognizant of that fact.

Powell. Just a practical comment about the matter of money. I guess what I am calling for is for more of us to go to molecular biologists and convince them that the questions we're asking and trying to solve are important. We can harvest their expertise and exploit their already well-equipped laboratories, rather than duplicating their facilities everytime a new technique comes up.

Koehn. It's nevertheless expensive, getting into the facilities.

Powell. Yes, sure, it's a little more expensive. I would still like to make a plea of trying to get us to sell ourselves to molecular biologists and have them do some of the work.

III. ....And What Shall We Measure It With?

Ideally, research starts with the posing of a question. If the question is to be answered with empirical methods the investigator must next consider the organism and environment to be used in the study. Obviously, research results may be strongly influenced by the choices that are made, and the generalization of the investigator's results to a broader domain may be severely limited if either pathological organisms or environments are used. The following interchange was provoked by Doug Gill, who eloquently expressed the reservations to laboratory research held by many field-oriented population biologists.

Gill. One of the features of this symposium that has surprised me is the inordinate interest that we have all expressed in the heritability of fitness components and the idea that this should be the principal line of future research. And yet the thing that has been strikingly lacking, in my judgment, has been attention to the fact that fitness measures are only relevant to the environment they are measured in. For instance, if you measure r it changes with every environment. We need to put our systems back into their natural environments so that relevant fitness measures are obtainable. I find myself increasingly intolerant or bored by measures of the genetic components of fitness of *Drosophila* in *Drosophila* bottles. Even though *Drosophila* is well known genetically, we know absolutely nothing about its natural ecology or population biology.

Dawson. That is not true—we know a lot about what is going on out there.

Gill. Even if it's not true, that's still no excuse for worrying and fussing about how slight changes in the medium will or will not influence fecundity measures of *Drosophila pseudoobscura*. Just look at the *Tribolium* experiments that Pete [Dawson] talked about.

Dawson. But Doug, you have become blinded by trees and ponds. The environments that we used as "selection regimes" in the experiments I discussed yesterday were carefully chosen because they in fact do represent exactly the sorts of environments in which the animals are found in nature.

Gill. Indeed! Natural environments: grainaries in Iowa. I think there is an extraordinary lack of attention to what constitutes the environment in which all this genetic machinery is relevant. Now I am interested in Dick Koehn's work because it has a semblance of reality in trying to take the LAP locus in mussels in Long Island Sound and make it relevant to the environment of interest to mussels. Paul Hebert's work is similar in trying to draw some relationship between the genes of the cladocerans and the relevant environments. My own work is motivated in the same direction. So to turn to future research, I'm increasingly disappointed with researchers who turn out laboratory experiments that in large measure have nothing to do with anything.

Roundtable Discussion 215

As an evolutionary ecologist I'm really not interested in the heritability of egg production in chickens.

Dawson. Are you interested in the heritability of egg production in salamanders?

Gill. If heritability accounts for such a low percentage of the variance in egg production as Connie [Istock] suggested in his talk and my student Keith Berven is finding in wood frogs, no, I'm not sure that I am. My job is to account for the 95 percent that is not explained by heritability. Consider, for example, diets or many physiological aspects of metabolism. I'd be willing to bet that most of those factors are relatively important to the organism.

Koehn. Yes, but if you would do some physiology, you wouldn't have to bet. You could decide on the basis of information.

Gill. To come back to my point, I would like to see NSF fund a set of *Drosophila* people to get out in the field, rather like Hampton Carson, and try to understand something about *Drosophila* in their natural environment. Less attention, or less funding, or whatever, should be given to laboratory characterization of *Drosophila*. As an ecologist I'm not interested in *Drosophila* in milk bottles in laboratories; as a geneticist I might be.

Audience. The work that has been done with *Drosophila* hasn't been done because people are interested in what *Drosophila* does in nature. It is because people have tried to use them to explore general problems of how organisms will evolve or respond to certain conditions and also because they can be manipulated in the lab. Whether *Drosophila* does something out in nature hasn't really been relevant for these types of experiments.

Gill. Yes, that's true. However, if we are interested in characterizing fitness components, if we're concerned about how natural selection works, what the relevance of genetic variation is, what controls population dynamics, or what the genetic relationships of interacting populations are, then we must study the setting in which all this takes place. That setting is the natural environment from the point of view of the organism. *Drosophila* in bottles are weird critters.

Audience. I disagree. What better way is there to go about exploring these questions than to give the organism novel experiments in which you can accurately observe responses. If you go out in nature, then you constrain the accuracy by the way you make the actual observation.

Hebert. But why should you adopt a second-best strategy? Why shouldn't you study an organism with which you can do both laboratory manipulations and field studies? I don't think Doug Gill has suggested that you give up experimental analysis, rather that you should be able to do them both.

Audience. Sure, I would agree doing both is also the best of all possible worlds. But there is, I think, a clear point to experimental work and that is to obtain information which, in lots of cases, just can't be gotten in the field.

Istock. I agree with our colleague in the audience that it depends on what question you are asking. I appreciate Gill's point, but it's not sufficient. I want to know the substructure of fitness all the way from the primary gene expression on up. It's only been possible to approach that problem in the laboratory environment where you can run carefully controlled experiments and tease apart components of fitness associated with morphological traits. That can be done with *Drosophila*. Sure, we're studying really funny organisms, flies in fly bottles that are peppered with mutant markers that would never make it in nature. But we're not trying to learn about ecology at that point; we're trying to learn something about the level or organization just below that the organism faces in the environment. If we really understood that level, we might be able to predict specific forms taken by phenomena that we are trying to understand in nature.

IV. Theory vs. Empiricism in Population Biology

Gill. I'd like to pose a question to Wyatt [Anderson] regarding his point about the explosive development of theory relative to empirical findings. You can actively do something about that, Wyatt, since you are on the NSF panel. Would you recommend that NSF divert its funding from theory in order to slow down its development and encourage more funding to go to experimental, perhaps even field studies?

Anderson. No, I don't think we should discourage theory. Theory is useful and I haven't meant to suggest that it is not. My point is that the field suffers from an imbalance between the two areas. We have a distinguished theorist in the audience, Monty Slatkin from the University of Washington. What about your response to this?

Slatkin. I certainly don't think NSF should stop funding theory. Theory will play an important part in population biology in the next ten years and I don't think the lack of contact between theory and experiments is all the fault of theoreticians. It is certainly the case that theoreticians make a number of simplifying assumptions about the way various processes work, but I think it is also the case that experimentalists make simplifying assumptions. For example, it is very nice to look at parthenogenetic systems to examine certain sorts of interactions, but the value of that has to be based on the assumption that those species represent the sexual species as well. A parthenogenetic species is a model system in the same way that a set of equations is.

Istock. We sometimes encounter problems when we attempt to generalize in population biology and that's why I said be careful about thinking only in terms of simple systems. The genetics of bacteria and elephants are not synonymous. I think that the power of theory is really increasing rapidly in biology; its major function is to take on those problems that cannot be solved by empiricism because they take a terribly long time to grub through fact by fact. Instead, by a leap of intuition, you get a result that takes you to a distant point which you can then approach again empirically. It's the same problem that's found, for instance, when you can't measure the movement of a comet by running along beside it. You have to do other things in order to make measurements which show whether there is a congruence or an anomaly with predictions.

Anderson. If I may compromise, I think theory will play an important role in the continued development of population biology. What concerns me is that a great deal of theory is essentially decoupled from experiments and cannot be tested by experimentalists.

Istock. We don't want to put fetters on the theoreticians, do we?

Anderson. No, I don't want to put fetters on anybody, but if the field is to develop, then there has to be closer interfacing between theory and experiments.

King. But theory frequently tends to stimulate experimentation even when it is wrong or cannot be properly tested. For instance, Robert MacArthur probably stimulated more empirical research than anyone in the history of ecology, yet most of his theoretical constructs have toppled. This I think, is the unique role of theory; any area of science must ultimately have its feet based on the soil of empiricism.

Udovic. (from audience). I'd like to make one comment about the connection between theory and experiments. Dr. Anderson argued that many times the results of theoretical studies don't lead to predictions that can be tested, either in the laboratory or in the field. But one of the values of the heuristic model, as I see it, is to take an assumption, which may or may not have been previously validated, and to look at the logical consequences of that assumption. For example, consider models that look at frequency dependent selection and say "if fitness values are frequency dependent, then the following results should hold." I think these models are quite valuable even though we may not know yet how important frequency-dependent selection is in natural systems. The result, as I see it the, is that heuristic models can point out the need for testing certain very important assumptions upon which the models are based. So I would turn the argument around and say that instead of looking for testable predictions, in many instances we should look at what important assumptions the models are based on and think about testing those.

Audience. I don't want to speak directly on that, but I think my comment relates. As an ecologist willing in principle to try to think about genetics, I'd like to express a feeling I've been getting throughout this meeting. If you come in as an ecologist and you attempt to tackle a typical ecological problem like why is an organism one way in one place and another way in another place, the obvious starting place is to say, "it's adaptive." Then you go to a meeting and someone says "oh, you can't just say it's adaptive because genes are more complicated than that and don't necessarily succeed in optimizing." So you say, "well, that sounds reasonable," and you go read a book and if you are very ambitious, you try to construct a single allele model. And then you go to another meeting and someone points out that genetic effects tend to be oversimplified and says "well really, a single allele model is not the way we must go; how about a multiple allele model or something a bit more realistic?" You go away and maybe one percent of you come back ten years later with some knowledge of multiple allele models and try to build one of those, and it produces lots and lots of different results, many of which may not be predictable. At that stage, someone says, "well it all depends still on whether there is inbreeding going on and things like that." At that point, you pick up a strong feeling of being lost in a dark forest. You tend to throw up your hands and go back to studying the ecosystem. Hopefully you haven't lost too much time.

V. Prospects for the Future

The current state of population biology is revealed more clearly and objectively by examination of a series of papers, such as those presented by participants in this colloquium, than by any other means. However, one of the stated aims of the Oregon State University meeting was to assess future developments and accordingly roundtable participants were asked to briefly comment on the one area that they thought would be most important in the next ten years. The responses of Koehn (ecological physiology) and Powell (evolutionary molecular biology) were presented in Section II of the roundtable. Those of the other participants are presented in this section.

Gill. One exciting area that is just breaking through is the genetics of host-parasite relationships. I see it as a field that will lead to a great deal of synthesis and combine the features that Wyatt [Anderson] has already pointed out. There are two or three areas within that context that are of immediate interest. One, of course, is the role of secondary compounds in plants. They appear to be principal features determining population abundances, distributions, survivorship, and fecundity of insects and pathogens. I don't think we yet know much, if anything, about the genetic basis of the diversity of secondary compounds. I see room for a lot of exciting research on the genetics of secondary compounds as they relate to host-insect relationships.

I made reference yesterday to the business of transposable elements in genetic systems. Some recent papers report antigenic variation in organisms like trypanosomes. The genetic mechanisms by which endoparasites might be changing the antigens on their surfaces to escape the immune responses of their hosts appear to be the same kind of genetic mechanisms that the hosts are using to produce antibodies in the first place. Such studies point to an interfacing of genetic mechanisms of hosts and parasites which is of great significance.

The third area was introduced years ago in phytopathology. It can be traced to the work of Flor, Person, and Ellingboe on the gene-for-gene concept of interactions between host and parasite. The phytopathologists are making use of plant pathogens, such as wheat rusts or powdery mildews, with economically important plants, such as cereal grasses. The genes for resistance in the hosts appear to behave in simple Mendelian ratios, are codominants, and they occur as allelic series at a few loci controlling resistance. Yet, the parasites, such as wheat rust, consistently break through the prevailing resistance of the host, e.g. wheat. When they do, they possess genes for pathogenicity which also behave in a simple Mendelian fashion, are recessives, and are always nonallelic. They appear at brand new loci. This work is well developed in the plant pathology literature but, as I said earlier, most of my colleagues are not familiar with it. Here again is an interfacing of population genetics, evolution, and population ecology of enormous portent. The genetic control of the interaction between host and parasite, especially host survivorship and the abundance of the pathogen, beautifully combines population ecology and population genetics in a way that is breaking new, exciting ground.

Dawson. I think that one of the most important things to be done is to try to understand the genetic basis of life history traits by application of quantitative genetics. A clear lesson from the experiments Russ Riddle and I did with our *Tribolium* populations is that pleiotropic effects of genes expressed as negative correlations among fitness traits can prevent adaptation to environments of the sort encountered by natural populations. We cannot hope to understand this without large-scale genetic analyses involving several traits at a time. Such studies are particularly important if, as Wyatt suggested, we are truly interested in bringing behavior into population biology. Most behavior that is at all interesting is under polygenic control. So there is a crying need for more studies on model organisms that can shed light on the genetic structure of fitness and its effects on life history components.

Istock. If Pete [Dawson] brings in behavior, I'm going to bring back cytogenetics. If we can begin to map important phenotypic traits down to chromosomes or chromosome segments, we may be poised to understand a lot about how things work when molecular and developmental biologists get their acts straight. Right now these are very primitive sciences. If you think we have too much theory, their problem is they don't have any. They are sort of putting it together, like tinker toys, as they go along. There is not such thing, for example, as a predictive theory of physical biochemistry and so their bumbling along is going to take a long time and we can't wait for that. It seems to me that we need to look at chromosomes, or some other intermediate, and solid, manifestation of genetic structure. If we can do that we can begin to connect pieces.

Levin. The plant literature is replete with hand-waving about the liability of inbreeding and the consequences of hybridity. It seems to me that one of the really important problems is to determine the effects of inbreeding and hybridity by manipulating genetic background. Then let's put the experimental plants into natural populations and actually determine, through a variety of demographic observations, exactly what the consequences of inbreeding and hybridity are over a series of environments for a given species. It is important to obtain this information for a series of different species, in inbreeders and outbreeders, and in trees, shrubs, and herbs. If we focus on the field demonstration of inbreeding effects, we can perhaps achieve results that are more realistic than those from greenhouse trials.

King. From a very personal view, I feel that if population biology is to make maximum progress in the next ten years we must study rotifers and other cyclical parthenogens more frequently than has been done in the past. One of the things that has attracted me, and Paul Hebert also, to this group is the interplay between sexual and asexual reproduction. Replicates of the same genotype can be obtained in these organisms by developing clones which can then be tested in a large series of environments. We can also do the opposite tests. That is, sexual reproduction can be used to produce many different genotypes, which can then be examined in a single environment. Such studies can be undertaken to partition genetic and environmental components of fitness. They can lead to a better understanding of the adaptive norms and developmental potentialities of a given genotype than can be obtained in any other way. In addition, by using cyclical parthenogens in carefully designed experiments, we should be able to look at, in a very direct fashion, the interaction between the life history of an organism and the way it exploits the environment. Such studies will not only answer old problems, they will lead to new ones.

Hebert. Let me first say that to my mind there is remarkably little research in population biology being done. Most work continues to be close to pure population genetics or ecology. There has been little emphasis on the fusion of ecology and genetics. Ecologists are concerned about population dynamics, while geneticists concentrate on allele dynamics. Surely, a central problem for population biologists is to determine how shifts in gene frequencies influence population characteristics. To answer this question we need to know much more about the relationship between changes in the environment and changes in the fitness of genotypes. George Williams has stressed the notion of a sisyphean genotype whose fitness is strongly dependent on environmental conditions. High fitness in multiple environments is seen as unlikely. The hypothesis of r and K selection is another instance in which high fitness in multiple environments is viewed as impossible. In models such as these, covariation in the fitnesses of different genotypes is low. On the other hand Wyatt Anderson has discussed the possibility that genetic variation is only involved in the fine tuning of population responses—as would be the case if all genotypes had a strongly correlated response to environmental change. To distinguish between these alternatives I think we have to study organisms which reproduce clonally—so that the fitness of individual genotyes can be compared in an array of environments. Conrad Istock made a good point when he suggested that we think seriously about studying bacterial systems. Certainly Luckinbill's (*Amer. Nat.* [1979] 113:427–437) work on r and K selection in protozoans serves as a model for the sort of studies I've suggested. I would therefore like to amplify Charles King's call for the widespread use of eukaryotes reproducing by facultative parthenogenesis.

Anderson. If I had to pick a single area, I would like to understand the mechanism of natural selection better in a true population biological context. I would like to know how the environment impinges on the organism and how it brings about selection. The ultimate kind of study is what Dick Koehn and others are doing at a biochemical-physiological level, but the system I work with is different. I would like to know, in nature, how the chromosomes that I study change in frequency; whether selection is a process that goes on all the time, up and down; what makes it fluctuate; what in the environment controls the system; whether the life history stage affected most is fertility; and finally how all these elements affect the picture I have now of the role of viability and fertility. Ultimately, I would like to decompose selection into physiological and biochemical processes related to specific genes.

Dawson. Connie [Istock] brought up an interesting point in his comments on the future of population biology. It is a point that I planned to pose to Wyatt [Anderson] in the form of a question. Returning to the Syracuse Symposium on Population Biology, I recall that Dick Lewontin thought the three important fields to fuse in the synthesis would be genetics, ecology, and development. Wyatt, you argued in your introductory remarks for integrating behavior, rather than development, with ecology and genetics. I agree with you completely, although behavior has fallen into some bad times recently. My question is this. What has happened to development? Was Dick wrong? Or has the field changed in orientation since 1968?

Anderson. I'm not sure why development hasn't played a bigger role in population genetics. Perhaps it has been due to the difficulties of connecting developmental processes to the environment on one side and to the genetic system on the other. I believe development *is* entering population biology now, through studies of gene regulation and gene expression like those Jeff [Powell] mentioned. The distinction between genetics and development at this level is blurred, however, and many of these studies are labeled as genetics rather than development.

Author Index

Abraham, I., 105, 112
Adams, M. W., 178, 184
Ahmad, M., 119, 141
Aitken, T. H. G., 110, 112
Albrecht, L., 171, 184
Allard, R. W., 42, 57, 171, 177, 184, 185
Almond, R., 33
Andersen, V. L., 75, 76, 95, 150, 151, 153, 169
Anderson, W. W., 3, 34, 77, 89, 110, 112, 120, 145, 189, 195, 202, 203, 205, 208, 211, 217, 218, 220, 223, 224
Andjelkovic, M., 106, 114
Andrewartha, H. G., 3, 33
Angus, R. A., 40, 56
Antonovics, J., 177, 186
Arens, M. F., 118, 142
Arnheim, N., 120, 122, 141, 146
Arnold, J., 109, 114
Atchley, W. R., 49, 56
Avieli, E., 179, 186
Avise, J. C., 118, 144, 200, 202
Ayala, F., 78, 89, 100, 102, 103, 112, 119, 120, 141, 144, 145, 148, 165, 168, 192, 199, 202, 204

Bader, R. S., 71, 90
Bailey, T. B., 70, 94
Baird, R. H., 137, 141
Baker, H. G., 35
Banyard, B. J., 176, 184
Barnes, B. W., 64, 92, 93
Bateman, A. J., 181, 184
Bateson, W., 61
Bayne, B. L., 119, 120, 121, 122, 125, 134, 135, 139, 141, 142, 143, 145

Beardmore, J. A., 119, 141
Beaty, B. J., 110, 112
Bell, A. E., 75, 76, 91, 95, 150, 151, 153, 155, 158, 164, 166, 167, 168, 169, 170
Bellacera, P., 141
Beltran, I. C., 179, 184
Berger, E., 38, 39, 56, 102, 104, 112
Berisford, C. W., 110, 112
Bernstrom, P., 177, 184
Berven, K. A., 1, 18, 20, 21, 23, 24, 26, 28, 29, 33, 34, 191, 215
Bewley, G. C., 107, 113
Bijlsma, R., 118, 142
Bijlsma-Meeles, E., 118, 142
Birch, L. C., 3, 33, 34
Birley, A. J., 148, 168
Bishir, J., 177, 184
Bishop, S. H., 121, 142
Blair, P. V., 151, 155, 169
Blomstrom, G. H., 33
Boag, P. R., 63, 90
Böcher, T. W., 177, 184
Bocquet, C., 118, 142
Bodmer, W. F., 193, 202
Bossert, W. H., 1, 34
Boylan, W. J., 80, 90, 95, 151, 155, 166, 169
Braskerud, O., 155, 167
Breese, E. L., 78, 84, 85, 90
Broadhurst, P. L., 80, 92
Brown, A. H. D., 171, 177, 178, 184, 186, 187
Brown, A. J. L., 102, 104, 112
Brown, C. K., 77, 91
Brown, J. L., 34
Brugger, K., 33
Bryant, E., 1, 34

Bucio, A., 178, 185
Bulmer, M. G., 64, 90
Buroker, N. E., 120, 142

Campbell, I. M., 75, 90
Canvin, D. T., 180, 185
Caraco, T., 89
Carfagna, M., 118, 142
Carson, H. L., 191, 202, 215
Caspari, E., 89, 164, 166
Cavalli-Sforza, L. L., 193, 202
Chambers, G. K., 119, 145
Chapco, W., 81, 90
Chapman, A. B., 80, 90
Charlesworth, B., 3, 34, 189, 192, 202
Charnov, E. L., 1, 34
Chew, K. K., 120, 142
Chinnici, J. P., 108, 112
Christiansen, F. B., 192, 202, 203
Chu, T. N., 180, 185
Clarke, B., 118, 142, 198, 202
Clarke, J. M., 74, 90
Clausen, J., 2, 34
Clayton, G. A., 74, 95
Clegg, M. T., 171, 185
Cody, M. L., 45, 56
Cole, L. C., 1, 34
Coles, J. F., 176, 178, 185
Collins, J. P., 1, 36
Comstock, R. E., 80, 90, 95, 155, 167
Connor, E. F., 199, 202
Cornish-Bowden, A., 137, 142
Coyne, J., 99, 112, 114, 202
Craft, W. A., 80, 90
Craig, J. V., 80, 90
Crane, M. B., 181, 185
Crease, T. C., 39, 43, 45, 54, 57
Crick, F., 210
Crow, J. F., 46, 56, 68, 90
Crumpacker, D. W., 175, 185

Dadlani, H. V., 80, 90
Dare, P. J., 137, 142
Darling, D., 33
Darlington, C. D., 90
Darwin, C., 61
David, J. R., 118, 119, 142, 145
Dawson, P. S., 75, 76, 89, 90, 147, 148, 150, 153, 156, 157, 162, 164, 165, 167, 168, 169, 211, 214, 215, 221, 224
Day, T. H., 118, 142
Dayton, D. E., 175, 185
Dean, R. T., 122, 142
Deaton, L. E., 137, 143
Delaca, T. E., 199, 202
Dempster, E. R., 147, 167
DePew, F., 141
Diamond, J. M., 45, 56
Dingle, H., 77, 91
Doane, W. W., 105, 112
Dobzhansky, Th., 61, 69, 77, 82, 91, 95, 100, 102, 104, 112, 195, 200, 203, 204
Downs, W. G., 110, 112
Doyle, R. W., 79, 91
Dudley, S. F., 110, 112
Dunn, L. C., 61, 95

Eanes, W. F., 100, 108, 113, 114
East, E. M., 62, 91
Edwards, D. B., 137, 142
Edwards, Y. H., 108, 112
El-Helw, M. R., 71, 96
Ellingboe, A. H., 220
Emlen, S. T., 34
Endler, J. A., 117, 142
Enfield, F. D., 152, 155, 167, 168
Englert, D. C., 75, 91, 150, 158, 167
Etges, W. J., 89
Evans, G., 181, 188
Ewing, E. P., 117, 121, 143

Fagen, R. M., 1, 34
Faini, D., 174, 185
Falconer, D. S., 71, 80, 81, 91, 149, 167
Falk, C. T., 104, 113
Feldman, M. W., 62, 91, 192, 202
Felsenstein, J., 148, 167
Felton, A., 99, 104, 114
Fenchel, T., 192, 203
Ferguson, A., 120, 145
Ferrari, C., 43, 57
Festing, W. M., 80, 93
Fieth, P., 138, 144
Finnerty, V., 99, 112
Fisher, R. A., 46, 57, 62, 67, 68, 69, 86, 91
Flor, H. H., 220
Fortuno, V., 179, 186

Author Index

Fouillet, P., 118, 142
Fowler, D. P., 176, 178, 185
Frankel, R., 185
Franklin, E. C., 175, 177, 185
Franklin, I., 197, 203
Fraser, A., 39, 59
Free, J. B., 181, 185
Fritz, R., 33
Fryxell, P. A., 181, 185
Fucci, L., 118, 142
Fulton, W. C., 73, 93
Fundiller, D. L., 120, 144
Futuyma, D. J., 45, 47, 58

Gabbott, P. A., 137, 143
Gadgil, M., 1, 34
Galderman, H., 67, 91
Gall, G. A. E., 151, 167
Gall, L. F., 198, 203
Galton, F., 62
Galun, E., 185
Gaso, M. I., 195, 202
Gaudio, L., 118, 142
Gianola, D., 70, 91
Gibson, J. B., 118, 143
Gilbert, W., 107, 112
Gill, D. E., 1, 4, 7, 9, 11, 20, 21, 23, 34, 191, 213, 214, 215, 216, 217, 220
Ginevan, M. R., 117, 121, 143
Ginzburg, L., 100, 113
Gorla, M. S., 174, 185
Gould, S. J., 196, 203
Gowen, J. W., 81, 91
Graham, J., 89
Graham, S. M., 198, 203
Grant, P. R., 63, 90
Grant, V. 177, 185
Griffing, B., 178, 186
Gruenstein, E. I., 102, 114
Guzman, J., 195, 202

Haines, P., 141
Hairston, N. G., 1, 35, 87, 91
Hall, J., 141
Halverson, T., 18, 19, 33
Hamilton, W. D., 196, 203
Hann, B. J., 39, 57
Hardin, R. T., 151, 152, 153, 164, 167, 168
Hare, J. D., 45, 47, 58

Harper, J. L., 178, 185
Harris, D., 70, 91
Harris, H., 108, 112, 147, 167
Harris, R., 33
Harrison, B. J., 73, 80, 93
Hayes, M. B., 100, 113
Hazel, J. R., 136, 143
Hebert, P. D. N., 37, 39, 40, 43, 45, 53, 54, 57, 119, 146, 214, 216, 222, 223
Hedgecock, D., 120, 141
Hedrick, P. W., 117, 121, 143, 167
Hegmann, J. P., 77, 91
Hermesen, J. G. T., 180, 185
Hershberger, W. K., 120, 142
Herzog, G. A., 77, 92
Heslop-Harrison, J., 189, 203
Hiesey, W. M., 2, 34
Hilbish, T. J., 137, 141, 143
Hill, J., 178, 185
Hill, W. G., 70, 92
Hillier, P. C., 118, 142
Hoekstra, R., 148, 168
Hollingsworth, M. J., 72, 73, 92
Hopkinson, D. A., 108, 112
Horovitz, D. G., 132, 144
Hoy, M. A., 77, 93
Hubby, J. L., 98, 113, 147, 168
Hunte, W., 79, 91
Hutchinson, G. E., 43, 45, 57

Immermann, F., 120, 137, 139, 141, 143
Innes, D., 141
Istock, C. A., 2, 61, 64, 69, 70, 71, 76, 82, 86, 87, 92, 93, 193, 207, 209, 215, 216, 217, 218, 221, 223, 224

Jaenicke, J., 47, 55, 57, 89
Jain, S. K., 42, 57, 177, 184
James, S. H., 176, 179, 184
Janossy, A., 177, 185
Janzen, D. H., 43, 58
Jarvinen, O., 47, 59
Jinks, J. L., 70, 80, 92, 93
Johnson, C. M., 174, 186
Johnson, E. C., 179, 186
Johnson, F. M., 118, 146
Johnson, G., 99, 112
Jones, D. F., 178, 186
Jones, P. J., 79, 82, 94

Kahler, A. L., 171, 184, 185
Katz, A. J., 152, 155, 167
Kaufman, P. K., 155, 167
Kearsey, M. J., 63, 64, 80, 92, 93
Keck, D. D., 2, 34
Kempthorne, E., 70, 94
Kernick, M. D., 181, 186
Kerster, H. W., 171, 172, 179, 181, 186
Kidwell, J. F., 79, 93
Kidwell, M. G., 108, 113
Kimmich, R. H., 110, 112
Kimura, M., 38, 46, 49, 58, 68, 90
King, C. E., 3, 35, 162, 168, 195, 202, 203, 207, 218, 222, 223
King, T. P., 100, 114
Kitty, C., 33
Koehler, R., 100, 113
Koehn, R. K., 108, 113, 117, 118, 120, 121, 122, 123, 124, 125, 130, 133, 134, 135, 136, 137, 139, 142, 143, 144, 145, 146, 198, 206, 209, 210, 211, 212, 213, 214, 215, 220, 224
Kojima, Ken-Ichi, 63, 80, 81, 92, 95
Koski, V., 177, 186
Krause, E., 153, 155, 168
Kyle, W. H., 151, 155, 169

Lack, D., 1, 34, 35
Lakatos, I., 197, 203
Lakovaara, S., 104, 113
Lande, R., 64, 65, 66, 69, 70, 85, 92
Lane, P. A., 45, 58
Langley, C. H., 102, 104, 112, 114
Lansman, R. A., 200, 202
Lassen, H. H., 120, 123, 124, 127, 130, 132, 144
Latter, D. H., 178, 186
Laurell, C. B., 134, 144
Laurie-Ahlberg, C. C., 107, 113
Leigh, E. G., 49, 58
Lent, C. M., 137, 144
Leon, J. A., 194, 203
Lerner, I. M., 149, 168
Levene, H., 147, 148, 168, 189, 192, 194, 203
Levin, D. A., 171, 172, 173, 179, 181, 186, 206, 208, 209, 210, 222
Levine, L., 195, 202
Levins, R., 147, 165, 168, 189, 193, 196, 203, 211
Levinton, J. S., 119, 120, 127, 144
Lewis, H., 177, 186

Lewontin, R. C., 1, 35, 38, 40, 42, 58, 62, 67, 69, 77, 91, 93, 97, 98, 99, 100, 104, 113, 114, 121, 144, 147, 168, 189, 191, 196, 197, 203, 206, 224
Li, C. C., 120, 130, 131, 132, 144
Lichtenfels, J. M., 105, 114
Likens, G. E., 58
Linney, R. B., 64, 93
Livingston, D. R., 138, 144
Lonnquist, J. H., 179, 186
Lopez-Fanjul, C., 150, 169
Lucchesi, J. C., 107, 113
Luckinbill, L. S., 223
Lumme, J., 77, 94
Lupton, F. G. H., 177, 185
Lynch, C. B., 77, 93

MacArthur, R. H., 3, 35, 44, 58, 194, 199, 203, 218
McCarron, M., 99, 112
McConkey, E. H., 102, 104, 113
McDonald, J. F., 118, 119, 144, 145, 148, 168, 192, 204
McNeilly, T., 177, 186
McVetty, P. B. E., 180, 185
McWilliam, J. R., 178, 186
Maguire, L. A., 191, 204
Maisel, E., 89
Makarewicz, J. C., 45, 58
Manning, A., 63, 93
Marani, A., 179, 186
Marien, D., 74, 93
Marinkovic, D., 165, 168
Maroni, G., 107, 113
Marshall, D. R., 171, 184
Marshall, K., 89
Mather, K., 66, 70, 73, 78, 80, 83, 84, 85, 90, 93, 181, 185
Mathison, M. J., 178, 186
Matzinger, D. F., 179, 186
May, R. M., 3, 35, 199
Maynard Smith, J., 3, 35, 46, 49, 51, 58, 69, 72, 73, 74, 90, 92, 93, 148, 168, 196, 204
Mayr, E., 47, 58
Mendel, G., 61
Merkle, L., 89
Merril, C. R., 103, 114
Merritt, R. B., 118, 145
Meyer, H. H., 152, 168
Miklos, G. L. G., 107, 115

Author Index

Milkman, R., 100, 113, 120, 123, 124, 130, 136, 143, 145
Minawa, A., 148, 168
Miracle, M. R., 45, 58
Mitter, C., 45, 47, 58
Mitton, J. B., 45, 59, 120, 123, 124, 136, 143
Mock, B. A., 1, 7, 9, 11, 33, 191
Moeur, J. E., 76, 89, 93
Molineaux, C., 33
Moll, R. H., 179, 186
Moore, C. H., 151, 166
Moore, D. M., 177, 186
Moore, M. N., 120, 121, 122, 125, 134, 135, 141, 142, 143, 145
Moran, C., 40, 57
Moran, G. F., 178, 186
Moran, P. A. P., 196, 204
Morris, R. F., 73, 82, 85, 93
Mortimore, G. E., 122, 145, 146
Mukai, T., 75, 86, 93
Mulcahy, D. L., 174, 186
Muller, H. J., 46, 58
Munstermann, L. E., 109, 114
Murakami, K., 174, 186
Murdock, E. A., 120, 145
Murphy, G. I., 1, 35

Nash, D. J., 79, 93
Nei, M., 78, 96
Nettancourt, D. de, 173, 187
Nevo, E., 98, 113, 148, 168
Newell, C., 141
Newell, R. I. E., 120, 137, 139, 141, 143
Nicholas, F. W., 70, 92
Nilson, L. R., 118, 145
Nilsson, B. D., 179, 187
Nilsson-Ehle, H., 62, 93
Nordskog, A. W., 63, 80, 93
Norman, R., 89
Norton, D., 33
Nur, U., 89

O'Donald, P., 63, 70, 94
O'Farrell, P. H., 102, 113
Ohnishi, O., 78, 94
Ohnishi, S., 77, 78, 96
Ohta, T., 38, 49, 58
Oikarinen, A., 77, 94
Oka, H. I., 180, 185, 187
Okada, I., 151, 152, 153, 168

Olvera, O., 195, 202
Oring, L. W., 1, 34
Orozco, F., 150, 153, 158, 169
Orr-Ewing, A. L., 179, 187
Ottaviano, E., 174, 185
Owen, G., 122, 145
Owino, F., 179, 187

Pandey, K. K., 175, 187
Park, Y. C., 70, 94
Parker, E. D., 45, 58
Parker, W. W., 1, 35
Parsons, P. A., 62, 63, 94
Paterniani, E., 183, 187
Pederson, D. G., 178, 187
Pepper, W. D., 177, 184
Perrins, C. M., 79, 80, 82, 94
Person, C., 220
Pfahler, P. L., 174, 187
Phan, D., 102, 104, 113
Philips, J. R., 77, 92
Phillips, M. A., 177, 187
Pianka, E. R., 1, 3, 35, 193, 204
Place, A. R., 118, 119, 145
Platt, J. R., 199, 204
Plowright, R. C., 173, 187
Pollack, E., 70, 94
Pontin, R. M., 47, 58
Popper, K., 199, 204
Porter, J. W., 191, 204
Powell, J. R., 97, 98, 100, 102, 104, 105, 106, 109, 112, 113, 114, 148, 162, 169, 192, 195, 202, 204, 205, 206, 207, 208, 210, 211, 213, 220, 224
Powers, D. A., 118, 119, 145
Prabhu, S. S., 80, 81, 90, 96
Prakash, S., 63, 82, 94
Preston, F. W., 43, 59
Price, M. V., 176, 178, 187
Prosser, C. L., 136, 143
Prout, T., 70, 74, 89, 94, 195, 200, 202, 204
Provine, W. B., 62, 95

Racine, R. R., 102, 114
Rahnefeld, G. W., 80, 95
Ramshaw, J. A. M., 99, 100, 114
Rao, R., 118, 142
Rasmuson, B., 118, 145
Rasmuson, M., 118, 145
Reeve, E. C. R., 81, 95

Rempel, W. E., 80, 90
Rich, E. R., 158, 169
Richards, A. J., 55, 59
Richardson, R. H., 81, 95
Rico, M., 100, 101, 102, 106, 114
Riddle, R. A., 147, 153, 155, 156, 163, 164, 165, 169, 170
Roberts, R. C., 79, 95
Robertson, F. W., 63, 80, 81, 95
Roche, L., 177, 187
Rohlf, F. J., 135, 145
Rosa, M. E. de la, 195, 202
Rothman, E. D., 196, 204
Roughgarden, J., 3, 35, 189, 192, 204
Ruano, R. G., 150, 169
Russell, E. S., 79, 95

Salceda, V. M., 77, 78, 95, 195, 202
Sang, J. H., 74, 95
Saura, A., 47, 59, 104, 113
Schaal, B. A., 120, 145, 172, 187
Schaffer, W. M., 1, 34, 35
Scheinberg, E., 75, 76, 95, 150, 151, 153, 169
Schneider, J. C., 45, 47, 58
Schworer, C. M., 122, 145
Seed, R., 120, 145
Segal, H. L., 122, 145
Selander, R. K., 47, 55, 57
Shade, R. O., 200, 202
Shank, D. B., 178, 184
Shifrin, S., 103, 114
Shope, R., 110, 112
Short, A. C., 183, 187
Sickel, J., 33
Siebenaller, J. F., 120, 133, 134, 135, 141, 143
Simberloff, D. S., 199, 202, 204
Singh, R., 99, 104, 114
Sinnott, E. W., 61, 95
Skibinski, D. O. S., 119, 141
Slatkin, M., 67, 95, 217
Smith, M. Y., 39, 59
Smith-Gill, S. J., 20, 21, 23, 33, 35
Sokal, R. R., 135, 145
Sokoloff, A., 149, 154, 169
Sondhi, K. C., 74, 90
Sorensen, F., 177, 187
Spano, A., 118, 142
Spencer, E. M., 100, 114
Spiess, E. B., 77, 95

Standbury, C., 33
Stearns, S. C., 1, 36, 65, 87, 96
Stebbins, G. L., 35
Stephenson, R. R., 137, 143
Stern, K., 177, 187
Strickberger, M. W., 78, 96
Strong, D. R., 199, 204
Styer, D., 102, 114
Suchanek, T. H., 120, 144
Suomalainen, E., 47, 59
Sutherland, J., 39
Switzer, R. C., 103, 114
Szyska, L. A., 199, 204

Tabachnick, W. J., 109, 114
Tait, W. M., 81, 96
Takayanagi, K., 174, 186
Tantawy, A. O., 71, 96
Tappel, A. L., 122, 145
Taylor, B. J., 102, 104, 113, 148
Taylor, C. E., 162, 169
Templeton, A. R., 196, 204
Theisen, B. F., 120, 145
Thoday, J. M., 67, 96
Thompson, R., 70, 96
Thompson, S., 33
Thomson, J. D., 173, 187
Tinkle, D. W., 1, 34, 36, 87, 91
Tobari, I., 78, 96
Tong, G., 33
Tuljapurkar, S., 100, 113
Turano, F. J., 120, 123, 124, 130, 132, 144

Udovic, D., 218

Valentine, J. W., 120, 141, 199, 202
Van Delden, W., 118, 142
Van der Meulen-Bruijns, C., 118, 142
Vanko, W., 33
Vavra, K. J., 70, 71, 76, 86, 87, 92
Vepsalainen, K., 47, 59
Verghese, M. W., 80, 93
Vetukhiv, M., 77, 96
Vickery, R. K., 179, 188
Vigue, C. L., 118, 146
Vrijenhoek, R. C., 47, 59

Waddington, C. H., 147, 169
Wallace, B., 77, 96, 105, 114
Walton, K. E., 102, 114

Author Index

Warburton, F. E., 71, 96
Ward, R., 118, 122, 146
Ward, R. D., 39, 57
Ward, W. F., 122, 146
Warnick, J., 89
Wasser, N. M., 176, 178, 187
Wasserman, S., 92
Watanabe, T. K., 77, 96
Watt, W. B., 108, 115, 198, 203, 204
Weber, L., 102, 104, 112
Weir, B. S., 107, 113
Weiss, V., 70, 96
Wellner, D., 100, 113
Wernsman, E. A., 179, 186
West, C., 33
White, M. J. D., 47, 59
Whittaker, R. H., 46, 59
Widdows, J., 138, 144
Wiens, J. A., 3, 36
Wilbur, H. M., 1, 34, 36, 87, 91, 194, 204
Williams, G. C., 45, 49, 59, 223
Williams, R. D., 181, 188
Williamson, M. H., 148, 169
Wilson, A., 105, 115
Wilson, E. O., 3, 35

Wilson, S. P., 151, 155, 169
Wistrand, H. E., 148, 169, 192, 204
Woessner, R. A., 179, 188
Wong, W. C., 151, 155, 166, 169
Workman, P. L., 177, 184
Wright, M., 33
Wright, S., 68, 96, 164, 170, 182, 183, 188, 195, 196, 200, 204

Yamada, M., 174, 186
Yamada, Y., 151, 155, 164, 170
Yamamoto, M., 107, 115
Yamazaki, T., 75, 86, 93
Young, J. P. W., 39, 59, 120, 122, 146

Zeppezauer, E., 118, 145
Zera, A., 141
Zimmer, H., 92
Zinter, C., 33
Zirkle, D. F., 156, 157, 163, 164, 165, 166, 169, 170
Zisfein, J., 70, 71, 76, 86, 87, 92
Zobel, B., 179
Zumwalt, G. S., 120, 141, 199, 202

Subject Index

Adaptive landscape, 163
Altruism, 197

Behavior, 221
Behavioral ecology, 196, 197
Body size, 23, 28, 63, 72, 84
Breeding structure, 171, 182

Cannibalism, 76
Catalytic efficiency, 118, 135
Clinal variation, 20, 120
Clutch size, 24, 27, 28, 79
Coadaptation, 176
Coevolution, 189
Competition, 44, 45, 46, 63, 190, 191, 194, 195
Courtship patterns, 84

"Dead genes," 109
Density regulation, 3, 87
Development time, 2, 23, 24, 62, 64, 72, 73, 86, 87, 88, 89, 158
Diapause, 64, 77
Differential mortality, 133, 137
Dispersal, 53, 55, 84, 123, 139, 171, 172, 175, 210, 211
Dominance effect, 85

Ecotypes, 177
Environmental effects, 2
Environmental heterogeneity, 69, 148, 149, 211
Epistasis, 166
Evolutionary stable strategies, 196

Fertility, 62, 69, 79, 87, 88, 192
Fisher's theorem of natural selection, 82, 196
Fitness characters, 62, 64, 82, 161, 200
Fitness set, 196
Foraging efficiency, 63
Founder effect, 42, 100, 190

Gene flow, 123, 127, 172
Gene regulation, 105, 107
Gene selection vs. population selection, 127
Genetic differentiation, 109
Genetic distance, 49
Genetic drift, 163, 180, 181, 190, 191
Genetic load, 69, 177
Genetic variance, 2, 37, 51, 62, 69, 149, 163, 215
Genotype-environment interactions, 55, 56, 149
Geographical differentiation, 2, 206, 207

Habitat selection, 84, 162
Hardy-Weinberg equilibrium, 38, 39, 52, 53, 130, 132, 173
Heritability, 2, 62, 80, 81, 85, 213
Heterosis, 37, 38, 41, 43, 77, 179, 180, 206, 207
Heterozygote deficiency, 130, 131, 173
    superiority, 77
Hopeful monsters, 64
Host-parasite relationships, 220
Hybridization, 43, 175

Inbreeding, 41, 42, 72, 81, 164, 172, 173, 175, 176, 178, 182, 222

Subject Index

Inclusive fitness, 196
Intervening sequences, 107, 108
Isolation, reproductive, 81

Life history adaptations, 1, 2, 4, 18, 20, 30, 32, 65, 66, 86, 87, 88, 111, 189, 192, 194, 221
Linkage, 40, 73, 198
Longevity, 71, 79, 192

Macroevolution, 83
Maternal effects, 157
Mate selection, 84
Mating, assortative, 174
 random, 53
Meiosis, cost of, 49, 51, 53, 55, 56
Migration, 84
Mitochondrial DNA, 200
Molecular biology, 98, 111, 210, 212
Molecular evolution, 206
Mortality, 123, 129
Mutation, 48, 49, 52, 53, 66, 69, 108

Neighborhood size, 171, 183
Neutral alleles, 3, 38
Niche differentiation, 45, 55
Null alleles, 47, 48

Optimization theory, 109, 196
Osmoregulation, 122, 123
Outcrossing, 177
Overdominance, 165

Parasites, 4, 10, 11, 111, 220, 221
Parthenogenesis, 37, 45, 46, 111, 222, 223
Phenotypic variation, 62, 118
Pheromones, 84
Pleiotropy, 149, 164, 165, 221
Polygenic inheritance, 62, 63, 66, 78, 130, 149, 198, 221
Polymorphism, 2, 3, 41, 84, 117, 136, 140, 148, 192, 193, 195, 198
Population density, 15, 78

Population dynamics, 4, 43, 51, 88, 157, 190, 214, 216, 223
Population regulation, 190, 194
Post-transcriptional modification, 99, 100
Post-translational modification, 99, 100, 206
Preadaptation, 49
Predation, 63, 190
Primary fitness characters, 82, 88

r and K selection, 193, 223
Recombinant DNA, 200
Recombination, 41, 66, 68, 69, 87, 107, 108, 109
Ribosomal proteins, 102

Selection, artificial, 64, 74, 149, 164
 density-dependent, 193, 195
 directional, 62, 63, 66, 69, 84, 85, 86, 87, 89
 frequency-dependent, 219
 group, 46, 49
 neutralist controversy, 108
 stabilizing, 63, 64, 66, 75, 76, 86, 89
Selection coefficients, 40, 180
Selfishness, 197
Self incompatibility, 173, 175
Sex, evolution of, 108
Sexual dimorphism, 84
Sociobiology, 196, 197
Spatial variation, 120, 130
Speciation, 46, 82, 206
Species diversity, 44, 46
Substitutional load, 206
Syracuse Population Biology Colloquium, ix, 147, 189

Temporal variation, 117
Transposable elements, 220
Trypanosomiasis, 7, 8, 9, 10, 30, 220

Viability, 9, 62, 75, 77, 85, 87, 88, 157, 172, 178

Wahlund effect, 120, 130, 133
Wright's fixation index (F), 130, 131, 132

The Contributors

Wyatt A. Anderson. Department of Molecular and Population Genetics, University of Georgia, Athens, GA 30602.

Keith A. Berven. Department of Zoology, The University of Maryland, College Park, MD 20742.

Peter S. Dawson. Department of Zoology, Oregon State University, Corvallis, OR 97331.

Douglas E. Gill. Department of Zoology, The University of Maryland, College Park, MD 20742.

Paul D. N. Hebert. Department of Biology, University of Windsor, Windsor, Ontario N9B 3P4.

Conrad A. Istock. Department of Biology, University of Rochester, Rochester, NY 14627.

Charles E. King. Department of Zoology, Oregon State University, Corvallis, OR 97331.

Richard K. Koehn. Department of Ecology and Evolution, State University of New York, Stony Brook, NY 11794.

Donald A. Levin. Department of Botany, University of Texas, Austin, TX 78712.

Beverly A. Mock. Department of Zoology, The University of Maryland, College Park, MD 20742.

Jeffrey R. Powell. Department of Biology, Yale University, New Haven, CT 06520.

Russell A. Riddle. Department of Zoology, Oregon State University, Corvallis, OR 97331.

Bei Fragen zur Produktsicherheit wenden Sie sich bitte an:
If you have any questions regarding product safety,
please contact:

Walter de Gruyter GmbH
Genthiner Straße 13
10785 Berlin
productsafety@degruyterbrill.com